Developmental Plasticity of Inhibitory Circuitry

Sarah L. Pallas
Editor

Developmental Plasticity of Inhibitory Circuitry

Editor
Sarah L. Pallas
Georgia State University
Dept. Biology
24 Peachtree Center Ave.
428 Kell Hall
Atlanta, GA 30303
spallas@gsu.edu

ISBN 978-1-4419-1242-8   e-ISBN 978-1-4419-1243-5
DOI 10.1007/978-1-4419-1243-5
Springer New York Dordrecht Heidelberg London

Library of Congress Control Number: 2009932756

© Springer Science+Business Media, LLC 2010
All rights reserved. This work may not be translated or copied in whole or in part without the written permission of the publisher (Springer Science+Business Media, LLC, 233 Spring Street, New York, NY 10013, USA), except for brief excerpts in connection with reviews or scholarly analysis. Use in connection with any form of information storage and retrieval, electronic adaptation, computer software, or by similar or dissimilar methodology now known or hereafter developed is forbidden.
The use in this publication of trade names, trademarks, service marks, and similar terms, even if they are not identified as such, is not to be taken as an expression of opinion as to whether or not they are subject to proprietary rights.

Printed on acid-free paper

Springer is part of Springer Science+Business Media (www.springer.com)

Preface

This book was conceived at Watershed Restaurant in Decatur, GA, after a symposium on the topic organized by myself and Pete Wenner for the 2006 Society for Neuroscience meeting in Atlanta (Pallas et al. 2006). Our compliments to Chef Scott Peacock! We put together the symposium because of our own interest in this under-studied topic, relevant findings in our laboratories, and the fact that several mechanistic explanations for plasticity at inhibitory synapses had been uncovered by the invitees. Due in large part to the work of the contributors to the symposium and to this book, inhibitory plasticity is finally becoming widely recognized as a critical area for investigation. Increasing evidence supports an important role for inhibition in disease states, including epilepsy, schizophrenia, and autism spectrum disorders, and one of our aims in this book has been to bring together data from the synaptic and circuit levels of analysis with some of the clinical data in one volume. Some of the authors we invited were ultimately unavailable to contribute chapters, but we had the great good fortune to be able to add several others. As in any collection, however, there are many more investigators and studies that we would like to have included but could not due to lack of space. It has been our goal to provide the reader with a broad overview of mechanisms underlying inhibitory plasticity and of the systems in which it operates. We hope that this book will encourage further study of inhibitory plasticity by other investigators, and that further elucidation of the underlying mechanisms will lead to translational applications.

Atlanta, GA   Sarah L. Pallas

Contents

Background

1. Introduction 3

2. The Origins and Specification of Cortical Interneurons 13
   Asif Maroof and Stewart Anderson

3. Role of Spontaneous Activity in the Maturation
   of GABAergic Synapses in Embryonic Spinal Circuits 27
   Carlos E. Gonzalez-Islas and Peter Wenner

Systems

4. Regulation of Inhibitory Synapse Function in the Developing
   Auditory CNS 43
   Dan H. Sanes, Emma C. Sarro, Anne E. Takesian, Chiye Aoki,
   and Vibhakar C. Kotak

5. Developmental Plasticity of Inhibitory Receptive Field
   Properties in the Auditory and Visual Systems 71
   Khaleel A. Razak, Zoltan M. Fuzessery, and Sarah L. Pallas

6. Postnatal Maturation and Experience-Dependent Plasticity
   of Inhibitory Circuits in Barrel Cortex 91
   Qian-Quan Sun

Synaptic Mechanisms

7. GABAergic Transmission and Neuronal Network Events
   During Hippocampal Development 115
   Sampsa T. Sipilä and Kai K. Kaila

8 **Endocannabinoids and Inhibitory Synaptic Plasticity
   in Hippocampus and Cerebellum** 137
   Bradley E. Alger

Future Directions

9 **Interneuron Pathophysiologies: Paths to Neurodevelopmental
   Disorders** 167
   Kathie L. Eagleson, Elizabeth A.D. Hammock, and Pat Levitt

Index 185

Contributors

Bradley E. Alger
Department of Physiology, University Maryland School of Medicine,
655 W Baltimore Street, Baltimore, MD 21201, USA
balgerlab@gmail.com

Stewart A. Anderson
Department of Psychiatry, Neurology & Neuroscience, Weill Medical
College of Cornell, Box 244, Rm 929A Lasden, 1300 York Avenue,
New York, NY 10021, USA
SAA2007@med.cornell.edu

Chiye J. Aoki
Center for Neural Science, New York University, 4 Washington Place,
New York, NY 10003, USA
chiye@cns.nyu.edu

Kathie L. Eagleson
Keck School of Medicine, Zilkha Neurogenetic Institute, University of Southern
California, 1501 San Pablo Street, MC 2821, Los Angeles, CA 90089-2821, USA

Zoltan M. Fuzessery
Department of Zoology and Physiology, University of Wyoming,
Laramie, WY 82071-3166, USA
zmf@uwyo.edu

Carlos Gonzalez-Islas
Department of Physiology, Emory University, 615 Michael Street,
Atlanta, GA, USA
cegonza@physio.emory.edu

Elizabeth A.D. Hammock
Keck School of Medicine, Zilkha Neurogenetic Institute,
University of Southern California, 1501 San Pablo Street,
MC 2821, Los Angeles, CA 90089-2821, USA

Kai K. Kaila
Department of Bioscience, University of Helsinki, P.O. Box 65,
Viikinkaari 1 SF-00014 Helsinki 00014.Finland
kai.kaila@helsinki.fi

Vibhakar Kotak
Center for Neural Science, New York University, 4 Washington Place,
New York, NY 10003, USA
kotak@cns.nyu.edu

Pat Levitt
Keck School of Medicine, Zilkha Neurogenetic Institute, University of Southern
California, 1501 San Pablo Street, MC 2821, Los Angeles, CA 90089-2821, USA
plevitt@usc.edu

Sarah L. Pallas
Editor, Neuroscience Institute, Georgia State University,
38 Peachtree Center Avenue, Atlanta, GA 30303, USA
spallas@gsu.edu

Khaleel A. Razak
Department of Psychology, University of California-Riverside,
900 University Avenue, Riverside, CA 92521, USA
khaleel@ucr.edu

Dan H. Sanes
Center for Neural Science, New York University, 4 Washington Place,
New York, NY 100031, USA
sanes@cns.nyu.edu

Emma C. Sarro
Center for Neural Science, New York University, 4 Washington Place,
New York, NY 10003, USA
sarro@cns.nyu.edu

Sampsa T. Sipilä
Department of Clinical Neurophysiology,
Oulu University Hospital, FIN-90029 Oulu, Finland
Sampsa.Sipila@Helsinki.Fi

Qian-Quan Sun
Graduate Neuroscience Program, 1000 E. University Avenue,
Laramie, WY 82071, USA
neuron@uwyo.edu

Anne E. Takesian
Center for Neural Science, New York University, 4 Washington Place,
New York, NY 10003, USA
aet241@nyu.edu

Peter Wenner
Department of Physiology, Emory University, 615 Michael Street,
Atlanta, GA, USA
pwenner@physio.emory.edu

Part I
Background

Chapter 1
Introduction

1.1 Hemifield Neglect?

Neuroscience has long been focused on understanding neural plasticity and its regulation during development and in adulthood. Oddly, despite the known importance of inhibition in shaping neural responses, and the rich variety in subtypes of inhibitory neurons (see Chap. 2), experimental work in this area has focused almost entirely on plasticity at excitatory synapses. Now, that has changed and the gap in knowledge of inhibitory plasticity is rapidly being filled. A growing body of evidence suggests that plasticity at GABAergic and glycinergic synapses is of critical importance during both development and aging, and several mechanisms have been uncovered. The appearance of several excellent reviews on the topic (e.g. Akerman and Cline 2007; Maffei and Turrigiano 2008; Spolidoro et al. 2009) is a further indication that there is an increasing recognition of the importance of inhibitory plasticity. In this, and the chapters that follow, we provide a glimpse into some of the most salient findings in this long-neglected area of research.

Research on synaptic plasticity has been concentrated, in particular, on NMDA receptor-dependent, long-term potentiation (LTP) and depression (LTD) at excitatory synapses (see Bliss and Collingridge 1993; Malenka and Bear 2004; Massey and Bashir 2007; Yashiro and Philpot 2008, for review). The notion of use-dependent plasticity, as popularized by Donald Hebb, may explain the bias of the field toward excitatory synapses, and the neglect of the inhibitory side of plasticity. Hebb theorized that connections between neurons could get stronger if the postsynaptic neuron was successfully activated by the presynaptic terminal (Hebb 1949). Certainly, it was easy to envision how inhibitory connections could be turned down in strength by repeated use, simply because of their suppressive style, but it was not at all clear from this perspective how to get activation-based increases in synaptic efficacy at inhibitory synapses, whose usual function is to silence their targets. This is especially true because LTP/LTD involves calcium-dependent activation of signaling cascades, and the $GABA_A$ receptor is a Cl^- channel that does not pass calcium.

1.2 "Inhibition" is Excitatory Early in Development

A possible solution to the mystery became clear when it was discovered that GABA, the major inhibitory neurotransmitter in adult vertebrates, is actually excitatory at perinatal stages of development (Cherubini, 1991 #2074, see Ben-Ari 2002, for review). This paradoxical situation is due to immature levels of the KCC2 K^+-Cl^- transporter, resulting in a negative reversal potential for GABA channels, and thus chloride efflux on channel opening. Thus, plastic changes of many "inhibitory" connections between neurons could be limited to this period of time, and could work in the same way as connections that are excitatory throughout life. Evidence for this comes from reports that inhibitory GABAergic synapses could undergo LTP (called LTP-$GABA_A$) showed that calcium channel blockers could prevent it, pointing to a $GABA_A$ receptor-induced activation of voltage-dependent calcium channels (Caillard et al. 1999a). The plasticity was expressed by presynaptic changes in probablility of GABA release. These results then raised the question of how LTD of GABAergic synapses could occur. As a consequence, LTD of immature GABAergic synapses (called LTD-$GABA_A$) can occur by traditional means of depolarization-induced removal of the Mg^{++} block of NMDA receptors (Caillard et al. 1999b), which in turn, perhaps through a similar signal transduction pathway as LTD at excitatory synapses, can lead to reductions in presynaptic GABA release.

The work of the Wenner lab (Chap. 3) shows the importance of the excitatory action of immature GABA receptors, but on the other hand points out the need for homeostatic regulation of excitation to prevent epileptiform activity. GABARs that contribute to bouts of spontaneous network activity (SNA) in chick spinal cord *in ovo* that are critical for normal maturation of neural networks (see also Cancedda et al. 2007), and are tightly regulated through homeostatic mechanisms (see below) that seem to involve activity-dependent regulation of intracellular chloride levels.

1.3 Mechanisms of Inhibitory Plasticity are Highly Diverse

As the central nervous system matures, chloride transporters change their expression patterns such that NKCC1, which accumulates Cl^-, is down-regulated and KCC2, which exports Cl^-, is up-regulated, making the reversal potential for Cl^- more negative. As a result of this change in chloride balance, opening of the GABAR chloride channels becomes hyperpolarizing (Rivera et al. 1999; Lee et al. 2005). Through what mechanisms can inhibitory synaptic plasticity occur after this point? Several chapters illustrate the rich variety of ways in which this can occur.

Some of the earliest reports of changes in the strength of inhibitory synapses came from studies of brain regions in which inhibition was known to be an important contributor to behavioral output, such as the brainstem auditory areas involved in sound localization (e.g. Sanes and Rubel 1988; Sanes et al. 1992; Sanes and Takács 1993; Werthat et al. 2008, Chap. 4 in this volume). During development, along with the pruning of excitatory connections, GABAergic and glycinergic inhibitory synapses are pruned as well, helping to bring about an appropriate balance between inhibition and excitation in neural networks. After cochlear damage, inhibition is reduced, leading to increased excitability but a broadening of sound frequency tuning. A similar loss of inhibition occurs in age-related hearing loss (Caspary et al. 2008). The underlying mechanisms are diverse, and include both presynaptic and postsynaptic processes.

1.3.1 Co-Transmitters

Some inhibitory neurons, in addition to the release of their inhibitory neurotransmitter substances, also release excitatory neurotransmitters when they are activated, providing a sort of end run around the problem of facilitating inhibitory connections. In the MNTB, a form of LTD (LTDi) occurs at inhibitory synapses onto MSO neurons that involves postsynaptic $GABA_B$ receptors. Interestingly, these synapses contain glycine rather than GABA in adulthood. Even more surprising is the fact that in addition to GABA, immature MNTB neurons contain glutamate (Gillespie et al. 2005).

1.3.2 Changes in Receptor Subunit Composition

Another mechanism of inhibitory plasticity seen in the auditory pathway and elsewhere (Fagiolini et al. 2004) is an activity-dependent change in GABA receptor subunit composition. As with NMDA receptors (Stocca and Vicini 1998), there is a developmental progression in expression and incorporation of different GABA receptor subunits, providing an additional avenue for regulation (Golshani et al. 1997, see Chap. 6 for review).

1.3.3 DSI

In Chap. 8, Alger reviews how endocannabinoids can induce inhibitory plasticity (LTDi) in the hippocampus and cerebellum by presynaptic alteration of the strength of inhibitory synapses, in a process called depolarization-induced suppression of inhibition (DSI). Endocannabinoids (ecs) likely function as retrograde messengers (Chevaleyre et al. 2006, for review), and ec-LTD is triggered through mGluR-dependent release of ecs onto CB1 receptors at GABA terminals, depressing GABA release. In keeping with the depolarizing effect of GABA in neonates, ecs depress activity. Alger makes the important observation that cannabis use in pregnancy could have unintended consequences on brain development. A somewhat opposite process occurs via presynaptic NMDAR activation of GABAergic terminals that leads to LTPi of GABAergic synapses in Xenopus tectum (Lien et al. 2006).

1.3.4 Inhibitory STDP

Spike timing-dependent plasticity refers to the fact that some forms of Hebbian plasticity require that presynaptic activity evokes a spike in the postsynaptic neuron within a short time window of about 20 ms (Zhang et al. 1998). That this can also occur at inhibitory synapses was reported by Poo and colleagues (Woodin et al. 2003; Lu, 2007 #8849, reviewed in Caporale and Dan 2008). The underlying mechanism is activation of voltage-dependent Ca^{++} channels and a decrease in KCC2, reducing inhibition.

1.3.5 Receptor Trafficking

Activity-dependent regulation of receptor trafficking is a well-accepted explanation of LTP and LTD of glutamatergic synapses, but also occurs at inhibitory synapses (Marsden et al. 2007; Bannai et al. 2009). Excitatory activity can affect diffusion kinetics and receptor cluster size negatively, thus increasing susceptibility to LTP.

1.4 Homeostatic Plasticity

In retrospect, it may seem obvious that what goes up must come down, i.e. strengthening of excitatory synapses cannot go on indefinitely without reaching an inflexible maximum. A stable baseline is necessary for change to be recognized. Once the concept of homeostatic plasticity was introduced, the emphasis was on explaining how the strength of excitatory connections could be decreased after LTP. Homeostatic plasticity does involve inhibitory as well as excitatory plasticity, however, and can occur on several different levels. It was first described as a regulation of the basal activity setpoint of individual neurons in culture (Turrigiano et al. 1994; Turrigiano, 1995 #4671, see Davis 2006; Turrigiano 2007, for review). At synapses, the process is called synaptic scaling, and refers to a setpoint of synaptic strength that allows potentiation or depression to occur. Homeostasis can also occur at the network level, and functions to maintain flexibility in the face of input perturbations, including those resulting from loss of input. That homeostatic plasticity also occurs at inhibitory synapses has been demonstrated in our lab and others' (Razak and Pallas 2006; Razak and Pallas 2007; Carrasco, submitted #9364, reviewed in Chaps. 3, 5, 6, Nelson and Turrigiano 2008).

1.5 Critical Periods

Ocular dominance plasticity in visual cortex is perhaps the second most popular model behind hippocampus for studying mechanisms of neural plasticity, and the mechanism (NMDAR-dependent LTP/LTD) is much the same except that the extent to which excitatory plasticity can be evoked is dependent on age in cortex (reviewed in Malenka and Bear 2004; Smith et al. 2009). The additional involvement of inhibition in ocular dominance plasticity began to draw more attention after reports from Hensch and colleagues that mice with knockout of GAD65 (an isoform of glutamic acid decarboxylase, the synthetic enzyme for GABA) fail to exhibit LTD of connections from the closed eye (Hensch et al. 1998, see also Gandhi, 2008 #9156). Chap. 6 discusses how GAD can be regulated by experience, and thus promote maturation of inhibitory circuits in sensory cortex (see also Sun 2007).

1.6 Old Dogs and New Tricks: Adult Plasticity and Aging

In neuroscience in general, there is increasing recognition that, in some sense, critical periods never close but only fade away, with synaptic plasticity requiring more vigorous or prolonged stimulation with age (e.g. Linkenhoker and Knudsen 2002; Hofer et al. 2006; He et al. 2007; Zhou and Merzenich 2007; Spolidoro et al. 2009). One wonders why it has taken so long to realize that we old folks can still learn! The rescue of plasticity by benzodiazepines in GAD −/− mice suggests that it may be possible to reopen critical periods in adulthood through activity-dependent inhibitory plasticity (reviewed in Morishita and Hensch 2008). Indeed, there is accumulating evidence that inhibitory plasticity is especially important and common in adulthood. The impression that plasticity occurs mainly in juveniles may come primarily from an overgeneralization based on ocular dominance plasticity research. In visual cortex, sensory deprivation by monocular lid suture or dark-rearing was reported to have negative effects in juveniles but not adults, and these effects are long-lasting to the point of irreversibility (reviewed in Daw 1994), but this hard line view is softening. Research in my lab shows that superior colliculus can remain sensitive to dark-rearing long past critical period "closure" in cortex (Carrasco et al. 2005; Carrasco and Pallas 2006) as a result of inhibitory plasticity (Carrasco et al., submitted).

In addition to modifying existing synapses, in some parts of the adult brain including olfactory pathways, hippocampus, and cerebellum, entirely new neurons are produced and then integrated into existing circuits. This is also true of transplanted stem cells (Snyder et al. 1997). It would seem that these neurons face formidable obstacles by trying to differentiate into a circuit that has passed its malleable period. Fairly recently, however, it has come to light that GABA in adult hippocampal neurogenesis is depolarizing. Newborn granule cells in adult cerebellum go through the same change in Cl⁻ transporter expression pattern as in immature cerebellum. Their integration into a circuit is GABA-dependent, and once integrated they exhibit adult transporter expression patterns (Ge et al. 2007, 2008). In fact, GABA excitation normally occurs in adult brain in both transient and sustained modes (Marty and Llano 2005), and tetanic stimulation of GABAergic synapses in hippocampus can change Cl⁻ balance in a short or long-term way. Whether this is a physiologically relevant situation under normal conditions is an important question to address, but is likely to be relevant to generation of epileptic foci.

Inhibitory plasticity may even be more prevalent than excitatory plasticity in adult brain under some circumstances. Nedivi and colleagues find that in adult cerebral cortex inhibitory neurons are more likely to undergo structural modifications than are excitatory pyramidal neurons (Lee et al. 2006). This is not always a good thing, however. Inhibitory plasticity is a significant contributor to perceptual problems associated with aging. Baby boomers experiencing hearing loss en masse may not realize that much of the problem comes from a loss of inhibition, and that their detection thresholds may actually be improving (Caspary et al. 2008). A similar situation occurs in the visual system (Hua et al. 2006). These results suggest that treatment with GABA agonists may be more helpful than turning up the volume!

In the final chapter of the book, Levitt and colleagues present some clinical disorders thought to be related to problems in development and maintenance of inhibitory circuitry in cerebral cortex, including epilepsy, schizophrenia, autism spectrum disorders, and Fragile X syndrome. Here, although considerable progress has been made, much is still to be learned from studies of inhibitory plasticity that might be helpful in understanding and treating these diseases. For example, Kaila (Chap. 7, this volume, Vanhatalo and Kaila, submitted) points out that more thinking around the issue of neonatal seizures needs to be done, given that early GABAergic activity is important for maturation and that traditional GABA agonists are not likely to suppress seizures when GABARs are immature and excitatory.

1.7 Conclusions and Future Directions

Some common themes throughout the book are that inhibitory plasticity occurs in many different neuronal subtypes, at a diversity of CNS areas and through a plethora of distinct mechanisms. It is not possible or productive to consider only excitatory connections when trying to understand plasticity because circuits that produce behavior involve both excitatory and inhibitory elements. It is clear that one cannot assume that the action of GABA or glycine is inhibitory, and that either LTP or LTD can result from activation of GABAergic synapses.

So what is left to be done? Lessons have been learned from debates such as those that plagued the field of hippocampal LTP (Malenka and Bear 2004). Hopefully, they can be put to good use in dispassionately teasing out the separate and synergistic roles of each different type of inhibitory plasticity.

In the case of endocannabinoid-mediated plasticity, several details need to be worked out. What is the identity of the natural ligand of CB1Rs at specific synapses throughout the brain; is it the same everywhere under all circumstances? How does the endocannabinoid release process work, and is it a regulated step or does release occur inevitably as soon as endocannabinoids are synthesized? What are the natural physiological stimuli for endocannabinoid actions? Can variations in endocannabinoid regulation during the course of development alter neuronal wiring diagrams? Can cannabinoid receptor-dependent long-term synaptic depression be reversed? How can the endocannabinoid system best be exploited therapeutically for its anticonvulsant effects?

Further characterization of the structure and function of inhibitory neurons in situ is also needed, as well as studies of their role in producing behavior. There is surely a link between subtype of neuron and type of plasticity it can undergo or produce, independent of brain area. Likewise, there will be similarities and differences in mechanisms across brain regions. Most research has concentrated on $GABA_A$ receptors, with a fair amount of work on glycine receptors, but examining contributions of other receptor types to inhibitory plasticity would be valuable.

Some interesting remaining questions are, whether for any given time and place, inhibitory plasticity plays more of a role than excitatory plasticity, under what conditions each type of plasticity is evoked, and whether they can operate simultaneously or are in conflict. Sanes suggests that deprivation induced loss of inhibition preserves an early immature state. But this is likely not the case when plasticity is initiated in adulthood. Thus, an important question touched on by several of the authors is if and how inhibitory plasticity in adults differs from that during development. The availability of tools for labeling subpopulations of inhibitory neurons and specifically manipulating them (e.g. Lechner et al. 2002; Sugino et al. 2006; Wang et al. 2007; Zhang et al. 2007) will be a boon to future research in this area. Computational approaches may also help to point the way to testable hypotheses.

This book brings together the work of researchers investigating inhibitory plasticity at many levels of analysis and in several different preparations. This topic is of wide relevance across a number of different areas of research in neuroscience and neurology. Understanding mechanisms of adult plasticity has profound implications for clinical populations suffering from brain disorders (Ehninger et al. 2008). Medical problems such as epilepsy, mental illness, and movement disorders can result from malfunctioning inhibitory circuits. Further, the maturation of inhibitory circuits may trigger the onset of critical periods of increased neural circuit plasticity, raising the possibility that such plastic periods could be reactivated for medical benefit by manipulating inhibitory circuitry. It is therefore essential to understand how inhibitory connections can be altered. The time is ripe to review and synergize the present knowledge in this topic, in order to reconcile conflicting data and to promote further progress.

References

Akerman CJ, Cline HT (2007) Refining the roles of GABAergic signaling during neural circuit formation. Trends Neurosci 30:382–389

Bannai H, Levi S, Schweizer C, Inoue T, Launey T, Racine V, Sibarita JB, Mikoshiba K, Triller A (2009) Activity-dependent tuning of inhibitory neurotransmission based on GABAAR diffusion dynamics. Neuron 62:670–682

Ben-Ari Y (2002) Excitatory actions of GABA during development: the nature of the nurture. Nat Rev Neurosci 3:728–739

Bliss TVP, Collingridge GL (1993) A synaptic model of memory: long-term potentiation in the hippocampus. Nature 361:31–39

Caillard O, Ben-Ari Y, Gaiarsa J-L (1999a) Long-term potentiation of GABAergic synaptic transmission in neonatal rat hippocampus. J Physiol 518:109–119

Caillard O, Ben-Ari Y, Gaiarsa J-L (1999b) Mechanisms of induction and expression of long-term depression at GABAergic synapses in the neonatal rat hippocampus. J Neurosci 19:7568

Cancedda L, Fiumelli H, Chen K, M-m P (2007) Excitatory GABA Action Is Essential for Morphological Maturation of Cortical Neurons In Vivo. J Neurosci 27:5224–5235

Caporale N, Dan Y (2008) Spike timing-dependent plasticity: a Hebbian Learning Rule. Annu Rev Neurosci 31:25–46

Carrasco MM, Pallas SL (2006) Early visual experience prevents but cannot reverse deprivation-induced loss of refinement in adult superior colliculus. Vis Neurosci 23:845–852

Carrasco MM, Razak KA, Pallas SL (2005) Visual experience is necessary for maintenance but not development of receptive fields in superior colliculus. J Neurophysiol 94:1962–1970

Carrasco MM, Mao Y-T, Pallas SL (submitted) Inhibitory plasticity underlies visual deprivation-induced loss of retinocollicular map refinement in adulthood.

Caspary DM, Ling L, Turner JG, Hughes LF (2008) Inhibitory neurotransmission, plasticity and aging in the mammalian central auditory system. J Exp Biol 211:1781–1791

Chevaleyre V, Takahashi KA, Castillo PE (2006) Endocannabinoid-mediated syanptic plasticity in the CNS. Annu Rev Neurosci 29:37–76

Davis GW (2006) Homeostatic control of neural activity: From phenomenology to molecular design. Annu Rev Neurosci 29:307–323

Daw NW (1994) Mechanisms of plasticity in the visual cortex: The Friedenwald Lecture. Invest Ophthalmol Vis Sci 35:4168–4180

Ehninger D, Li W, Fox K, Stryker MP, Silva AJ (2008) Reversing neurodevelopmental disorders in adults. Neuron 60:950–960

Fagiolini M, Fritschy JM, Low K, Mohler H, Rudolph U, Hensch TK (2004) Specific GABAA circuits for visual cortical plasticity. Science 303:1681–1683

Ge S, Pradhan DA, Ming GL, Song H (2007) GABA sets the tempo for activity-dependent adult neurogenesis. Trends Neurosci 30:1–8

Ge S, Sailor KA, Ming GL, Song H (2008) Synaptic integration and plasticity of new neurons in the adult hippocampus. J Physiol 586:3759–3765

Gillespie DC, Kim G, Kandler K (2005) Inhibitory synapses in the developing auditory system are glutamatergic. Nat Neurosci 8:332–338

Golshani P, Truong H, Jones EG (1997) Developmental expression of $GABA_A$ receptor subunit and GAD genes in mouse somatosensory barrel cortex. J Comp Neurol 383:199–219

He H-Y, Ray B, Dennis K, Quinlan EM (2007) Experience-dependent recovery of vision following chronic deprivation amblyopia. Nat Neurosci 10:1134–1136

Hebb DO (1949) The Organization of Behavior. John Wiley, New York

Hensch T, Fagiolini M, Mataga N, Stryker M, Baekkeskov S, Kash S (1998) Local GABA circuit control of experience-dependent plasticity in developing visual cortex. Science 282:1504–1508

Hofer SB, MrsicFlogel TD, Bonhoeffer T, Hubener M (2006) Prior experience enhances plasticity in adult visual cortex. Am J Ophthalmol 141:1172

Hua T, Li X, He L, Zhou Y, Wang Y, Leventhal AG (2006) Functional degradation of visual cortical cells in old cats. Neurobiol Aging 27:155–162

Lechner HAE, Lein ES, Callaway EM (2002) A genetic method for selective and quickly reversible silencing of mammalian neurons. J Neurosci 22:5287–5290

Lee H, Chen CX, Liu YJ, Aizenman E, Kandler K (2005) KCC2 expression in immature rat cortical neurons is sufficient to switch the polarity of GABA responses. Eur J NeuroSci 21:2593–2599

Lee W, Huang H, Feng G, Sanes J, Brown E, So P, Nedivi E (2006) Dynamic remodeling of dendritic arbors in GABAergic interneurons of adult visual cortex. PLoS 4:e29

Lien CC, Mu Y, Vargas-Caballero M, Poo MM (2006) Visual stimuli-induced LTD of GABAergic synapses mediated by presynaptic NMDA receptors. Nat Neurosci 9:372–380

Linkenhoker BA, Knudsen EI (2002) Incremental training increases the plasticity of the auditory space map in adult barn owls. Nature 419:293–296

Maffei A, Turrigiano G (2008) The age of plasticity: developmental regulation of synaptic plasticity in neocortical microcircuits. Prog Brain Res 169:211–223

Malenka RC, Bear MF (2004) LTP and LTD: an embarrassment of riches. Neuron 44:5–21

Marsden KC, Beattie JB, Friedenthal J, Carroll RC (2007) NMDA receptor activation potentiates inhibitory transmission through GABA receptor-associated protein-dependent exocytosis of GABA(A) receptors. J Neurosci 27:14326–14337

Marty A, Llano I (2005) Excitatory effects of GABA in established brain networks. Trends Neurosci 28:284–289

Massey PV, Bashir ZI (2007) Long-term depression: multiple forms and implications for brain function. Trends Neurosci 30:176–184

Morishita H, Hensch TK (2008) Critical period revisited: impact on vision. Curr Opin Neurobiol 18:101–107

Nelson SB, Turrigiano GG (2008) Strength through diversity. Neuron 60:477–482

Pallas SL, Wenner P, Gonzalez-Islas C, Fagiolini M, Razak KA, Kim G, Sanes D, Roerig B (2006) Developmental plasticity of inhibitory circuitry. J Neurosci 26:10358–10361

Razak KA, Pallas SL (2006) Dark rearing reveals the mechanism underlying stimulus size tuning of superior colliculus neurons. Vis Neurosci 23:741–748

Razak KA, Pallas SL (2007) Inhibitory plasticity facilitates recovery of stimulus velocity tuning in the superior colliculus after chronic NMDA receptor blockade. J Neurosci 27:7275–7283

Rivera C, Voipio J, Payne JA, Ruusuvuori E, Lahtinen H, Lamsa K, Pirvola U, Saarma M, Kaila K (1999) The K^+/Cl^- co-transporter KCC2 renders GABA hyperpolarizing during neuronal maturation. Nature 397:251–255

Sanes DH, Rubel EW (1988) The ontogeny of inhibition and excitation in the gerbil lateral superior olive. J Neurosci 8:682–700

Sanes DH, Takács C (1993) Activity-dependent refinement of inhibitory connections. Eur J NeuroSci 5:570–574

Sanes DH, Markowitz S, Bernstein J, Wardlow J (1992) The influence of inhibitory afferents on the development of postsynaptic dendritic arbors. J Comp Neurol 321:637–644

Smith GB, Heynen AJ, Bear MF (2009) Bidirectional synaptic mechanisms of ocular dominance plasticity in visual cortex. Philos Trans R Soc Lond B Biol Sci 364:357–367

Snyder EY, Yoon C, Macklis JD (1997) Multipotent neural precursors can differentiate toward replacement of neurons undergoing targeted apoptotic degeneration in adult mouse neocortex. Proc Natl Acad Sci 94:11663

Spolidoro M, Sale A, Berardi N, Maffei L (2009) Plasticity in the adult brain: lessons from the visual system. Exp Brain Res 192:335–341

Stocca G, Vicini S (1998) Increased contribution of NR2A subunit to synaptic NMDA receptors in developing rat cortical neurons. J Physiol 507:13–24

Sugino K, Hempel CM, Miller MN, Hattox AM, Shapiro P, Wu C, Huang ZJ, Nelson SB (2006) Molecular taxonomy of major neuronal classes in the adult mouse forebrain. Nat Neurosci 9:99–107

Sun QQ (2007) The missing piece in the 'use it or lose it' puzzle: is inhibition regulated by activity or does it act on its own accord? Rev Neurosci 18:295–310

Turrigiano G (2007) Homeostatic signaling: the positive side of negative feedback. Curr Opin Neurobiol 17:318–324

Turrigiano G, Abbott LF, Marder E (1994) Activity-dependent changes in the intrinsic properties of cultured neurons. Science 264:974–977

Vanhatalo S, Kaila K (submitted) Emergence of spontaneous and evoked EEG activity in the human brain. In: The Newborn Brain: Neuroscience and Clinical Applications, 2 Edition (Lagercrantz H, Hanson M, Evrard P, Rod C, eds). Cambridge, UK: Cambridge University Press.

Wang H, Peca J, Matsuzaki M, Matsuzaki K, Noguchi J, Qiu L, Wang D, Zhang F, Boyden E, Deisseroth K, Kasai H, Hall WC, Feng G, Augustine GJ (2007) High-speed mapping of synaptic connectivity using photostimulation in Channelrhodopsin-2 transgenic mice. PNAS 104:8143–8148

Werthat F, Alexandrova O, Grothe B, Koch U (2008) Experience-dependent refinement of the inhibitory axons projecting to the medial superior olive. Developmental Neurobiology 68:1454–1462

Woodin MA, Ganguly K, Poo M (2003) Coincident pre- and postsynaptic activity modifies GABAergic synapses by postsynaptic changes in Cl^- transporter activity. Neuron 39:807–820

Yashiro K, Philpot BD (2008) Regulation of NMDA receptor subunit expression and its implications for LTD, LTP, and metaplasticity. Neuropharmacology 55:1081–1094

Zhang F, Aravanis AM, Adamantidis A, de Lecea L, Deisseroth K (2007) Circuit-breakers: optical technologies for probing neural signals and systems. Nat Rev Neurosci 8:577–581

Zhang LI, Tao HW, Holt CE, Harris WA, M-m P (1998) A critical window for cooperation and competition among developing retinotectal synapses. Nature 395:37–44

Zhou X, Merzenich MM (2007) Intensive training in adults refines A1 representations degraded in an early postnatal critical period. Proc Natl Acad Sci USA 104:15935–15940

Chapter 2
The Origins and Specification of Cortical Interneurons

Asif Maroof and Stewart Anderson

2.1 Introduction

The cerebral cortex is composed of neural networks that function through an intricate balance of excitation and inhibition. At the cellular level, these cortical networks consist of projection neurons and interneurons that primarily use the neurotransmitters glutamate and GABA, respectively. GABAergic interneurons make up 25–30% of the cortical neuronal milieu, and they play a vital role in modulating cortical output and plasticity (Whittington and Traub 2003; Wang et al. 2004). Cortical interneurons also play a role in regulating developmental processes in the forebrain, including neuronal proliferation and migration during the establishment of cortical circuitry (Owens and Kriegstein 2002; Hensch 2005).

Despite their prominent role in the function of the cortex, studies determining how interneuronal progenitors establish their specific fates have been relatively sparse. Cortical interneurons accomplish specific functions through a remarkable diversity of subtypes that vary in morphology, physiology, and neurochemical constituents (Monyer and Markram 2004). Because this diversity and the context-dependent maturation of interneuron-defining features appear after weeks of postnatal maturation, progress connecting the embryonic development of cortical interneurons to their differentiated fate has been slow. Consequently, little was known about the origin and molecular determination of cortical interneuron diversity until improved fate-mapping approaches and transgenic mice became available.

2.2 Origins of Cortical Interneurons

While cortical projection neurons derive from the dorsal (pallial) telencephalon and migrate along radial glia to their final laminar position in the cortical mantle zone, immunolabeling for GABA (DeDiego et al. 1994) and Dlx2 (Porteus et al. 1994)

A. Maroof and S. Anderson (✉)
Department Psychiatry Neurology and Neuroscience, Weill Cornell Medical College, 244,Rm 929A Lasden, 1300 York Ave, New York, NY, 10021, USA

reveals streams of progenitors migrating tangentially from the subpallium into limbic and cortical structures in the telencephalon. Analyses of Dlx1/Dlx2 mouse mutants, together with in vivo ablation experiments and co-labeling of migrating cells in slice culture experiments, suggested that this tangential migration consisted of interneuronal progeny (de Carlos et al. 1996; Tamamaki et al. 1997; Anderson et al., 1997; Parnavelas 2000; Marin and Rubenstein 2001). Tangential migrations of putative interneurons have been identified in several mammalian species, including mice (Anderson et al., 1997; Wichterle et al. 1999), rats (de Carlos et al. 1996; Lavdas et al. 1999), ferrets (Anderson et al. 2002b), and humans (Letinic et al. 2002; Wonders and Anderson 2006). In rodents and ferrets, the subpallium appears to be the primary source of cortical interneurons, whereas one study reported that in human embryos, most cortical interneurons undergo their terminal mitosis in the cortical subventricular zone (Letinic et al. 2002). We will address distinct regions within the telencephalon that have been implicated as potential origins for cortical interneurons, with particular emphasis on the neurochemically defined interneuron subgroups that those regions generate.

2.2.1 Medial Ganglionic Eminence

Although initial studies of interneuron tangential migration labeled cells within the lateral ganglionic eminence (LGE) (de Carlos et al. 1996; Tamamaki et al. 1997; Anderson et al., 1997), these studies did not establish whether these cells originated within the LGE itself or migrated to the LGE via other progenitor domains. Indeed, fluorescent dye labeling of the more ventrally located medial ganglionic eminence (MGE) revealed large streams of cells, most of which express GABA, migrating into the cortex (Lavdas et al. 1999). Mice lacking the homeobox transcription factor Nkx2.1 exhibit a complete loss of this migratory behavior and have a roughly 50% reduction of GABA+ cells in the neocortex just before birth (Sussel et al. 1999). Comparison of LGE and MGE-derived cells in vitro and in vivo showed that MGE cells retain a far greater propensity to migrate into the cortex (Wichterle et al. 1999; Anderson et al. 2001; Wichterle et al. 2001).

Due to the extensive timeframe for cortical interneurons to mature into distinct subgroups, taking several weeks in rodents, slice culture experiments proved to be inadequate for fate mapping studies. Subsequent experiments that involved transplanting genetically-labeled MGE progenitors in utero into the embryonic MGE (Wichterle et al. 2001; Butt et al. 2005), the lateral ventricle (Valcanis and Tan 2003), or in vitro onto a neonatal, cortical feeder layer (Xu et al. 2004), demonstrated that the majority of MGE-derived interneuronal progenitors went on to express either parvalbumin (PV) or somatostatin (SST). This expression defines two distinct neurochemical subgroups, along with their associated physiological characteristics and synaptic contacts, that together comprise roughly 60% of the cortical interneurons in mice and rats (Gonchar and Burkhalter 1997; Kawaguchi and Kubota 1997). Of particular interest, these studies rarely found MGE-derived interneurons that express calretinin (CR), a calcium binding protein that is largely non-overlapping with the SST or PV subgroups and primarily labels cells with a vertically oriented, bipolar or bitufted morphology (Rogers 1992; DeFelipe 1997), which suggests that most CR+ interneurons originate from a spatially or temporally distinct progenitor domain than those that express PV or SST. Taken together, these three neurochemically-defined subgroups make up approximately 80% of all interneurons within the cortex.

An MGE origin for most PV- or SST-expressing interneurons in mice has been further confirmed by genetic fate mapping studies (Fogarty et al. 2007; Xu et al. 2008). In addition, transplantation studies have begun to identify molecular subregions of the MGE that appear to be biased toward the generation of distinct interneuron groups. The Nkx6.2-expressing region of the most dorsal MGE appears biased toward the generation of SST-expressing cells, whereas the ventral two thirds of the MGE appears biased towards generating PV-expressing interneurons (Flames et al. 2007; Wonders et al. 2008). However, it is important to note that despite the tendencies for PV- or SST-expressing subgroups to have distinct physiological properties and patterns of axonal targeting, distinct MGE domains giving rise to distinct interneuron "types" (defined by combinations of neurochemical, physiological, and morphological characteristics) remain to be identified.

2.2.2 Caudal Ganglionic Eminence

In addition to the MGE, the caudal ganglionic eminence (CGE) is the other subpallial structure most strongly implicated in the generation of cortical interneurons (Anderson et al. 2001; Nery et al. 2002; Nery et al. 2003). Morphologically, the CGE exists as a fusion of the MGE and LGE beginning at the coronal level of the mid to caudal thalamus. The ventral CGE, like the MGE, expresses Nkx2.1, while the dorsal CGE strongly expresses *Gsh2* and *ER81,* two transcription factors that are required for the proper patterning of the LGE and olfactory bulb (Corbin et al. 2003).

Initial fate mapping experiments of the CGE at E13.5 in the mouse found that the CGE gives rise to deep-layer cortical interneurons, many of which express PV or SST, but not CR (Nery et al. 2002). This lack of CR+ cells may be due to the age of the telencephalic tissue, since nearly all CR+ interneurons undergo their final S-phase of the cell cycle after E14.5 (Xu et al. 2004). Indeed, selective dissection of the dorsal CGE at E14.5 gave rise to many CR+, bipolar cells after plating on a cortical feeder layer (Xu et al. 2004). In addition, in utero isochronic, homotopic transplants of E15.5 dorsal CGE primarily generated CR+ interneurons that exhibited distinct spiking characteristics indicative of that interneuron subgroup (Butt et al. 2005). Finally, explant cultures from GAD65-GFP transgenic mice also suggested that many cells migrating from the CGE to the cortex become vertically-oriented, CR-expressing interneurons (Lopez-Bendito et al. 2004).

Taken together, these experiments suggest that bipolar, vertically oriented CR-expressing interneurons are primarily generated within the Nkx2.1-negative region of the dorsal CGE. The ventral CGE, on the other hand, may generate PV- or SST-expressing interneurons, although the caudal migration of MGE-born progenitors through the CGE en route to the cortex is an equally plausible scenario (Butt et al. 2005; Yozu et al. 2005). A distinct subgroup of CR-expressing interneurons, that display multipolar morphologies and co-express SST (Xu et al. 2006), appear to originate from an Nkx2.1+ progenitor in the dorsal MGE or Nkx2.1+ domain of the ventral CGE (Xu et al. 2008).

2.2.3 Lateral Ganglionic Eminence

Although several studies have indicated that any LGE contribution to cortical interneurons is far smaller than that of the MGE (Wichterle et al. 1999; Anderson et al. 2001; Wichterle et al. 2001), evidence in support of the LGE as a source of interneurons bears mention. Although Nkx2.1 mutants lack a normal MGE domain, the cortex at E18.5 has only a 50% reduction of GABA-expressing cells (Sussel et al. 1999). While this could be attributable to an enhanced generation of CR+ cells from the dorsal CGE, the LGE-like region shows robust migration to the cortex at E15.5 in these mutants (Anderson et al. 2001; Nery et al. 2003). In addition, slice culture experiments where progenitors were labeled with the S-phase marker BrdU indicate that a small number of LGE-derived cells, some of which co-label for GABA, do migrate from the LGE to cortex (Anderson et al. 2001). Finally, explants taken from rat embryos, in which the MGE has been removed, continue to show robust migration from the LGE to the cortex, implying that the observed migration is not due simply to MGE cells migrating through the LGE (Jimenez et al. 2002). One possible explanation for these mixed results is the pleiotropic nature of the LGE, which consists of distinct progenitor domains along the dorsal-ventral axis that give rise to olfactory bulb interneurons and medium spiny striatal projection neurons (Stenman et al. 2003a). In addition, migration from the LGE to cortex has been shown to include oligodendrocytes after E14.5 (Kessaris et al. 2006). In sum, the current data support a minor contribution from the LGE to the cortical interneuron population, which does not seem to include the SST- or PV-expressing subgroups (Xu et al. 2004).

2.2.4 Rostral Migratory Stream

In contrast to cells from the LGE, cells taken from the rostral migratory stream (RMS) at birth can express CR when cultured on a cortical feeder (Xu et al. 2004). The relevance of this finding is difficult to assess because nearly all CR+ interneurons in P25 somatosensory cortex are born before E16.5 (Xu et al. 2004). Two potential scenarios to explain these findings are that cells may leave the RMS prior to reaching the olfactory bulb and instead migrate into the cortex, or simply that CR+ interneurons of the olfactory bulb exhibit the capacity to differentiate in an in vitro cortical environment. In support of the former model, immunohistochemical labeling for Dlx1, which labels migrating interneuron precursors within the RMS, also appears to label cells migrating from the RMS into the cortex (Anderson et al. 1999). Earlier migration from the rostral neuroepithelium of the lateral ventricle into layer I of the cortex has also been described for cells expressing CR, calbindin (CB), and GABA (Meyer et al. 1998; Zecevic and Rakic 2001; Ang et al. 2003). Taken together, these results suggest the possible involvement of the RMS in the generation of cortical interneurons expressing CR.

2.2.5 Septal Region

Another subpallial region that may contribute interneurons to the cerebral cortex is the septal area. Initial speculation that migrations from the septal region to the cortex may exist was made based on immunohistochemical labeling for Dlx1 (Anderson et al. 1999). More convincing evidence comes from the recent analysis of mouse mutants lacking the homeodomain-containing transcription factor *Vax1*, which is expressed in a pattern similar to *Dlx1* and *Dlx2* within the subcortical telencephalon (Taglialatela et al. 2004). At birth, *Vax1* mutants have a 30–44% reduction in GABA-expressing cortical neurons, with the greatest loss occurring within the rostral-most cortex. While the MGE is reduced in size, the septal region is almost completely absent in these mutants. Experiments conducted using slice cultures show cells migrating from the ventro-lateral septum into layer I of the rostral cortex, and this migration is lost in the *Vax1* mutants. These data, therefore, provide evidence for a septal contribution to the cortical interneuron population, although further experimentation is needed to definitively show this migration. These results may help explain the large-scale migration of later-born interneurons from layer I into the cortical plate (Ang et al. 2003; Hevner et al. 2004). Whether this migration represents distinct subtypes of interneurons, remains to be explored.

2.2.6 Cortex

Although several reports have shown that cultures of dorsal telencephalic progenitors have the capacity to generate GABAergic cells (Götz et al. 1995; He et al. 2001; Bellion et al. 2003; Gulacsi and Lillien 2003), very little evidence supports a cortical origin for interneurons in rodents (Xu et al. 2004). One study using mice expressing Cre under the Emx1 and Dbx1 promoters, two homeobox transcription factors exclusively expressed in pallial progenitors, found no colocalization of interneuron markers with Cre in adult cortical sections (Fogarty et al. 2007). This study indicated that the pallium does not give rise to interneurons in rodents. However, retroviral labeling of slice cultures from the human embryonic forebrain suggest that the majority of GABA+ interneurons in the human cortex originate from cycling progenitors in the cortical subventricular zone (Letinic et al. 2002). Although this intriguing finding has yet to be replicated in nonhuman primates, the observation that Nkx2.1, a gene required for the specification of the MGE-derived interneuron subgroups in mice, is strongly expressed in the cortical proliferative zone in humans but not in rodents (Rakic and Zecevic 2003). Interestingly, based on an apparent increase in the numbers of "neurons with short axons" in Golgi-stained sections from humans compared to nonprimate species, Cajal proposed that the enhanced cognitive abilities of humans has resulted from increased representation of these cells (DeFelipe and Jones 1988).

2.3 Birthdating of Cortical Interneurons

Through various morphogens that induce cell divisions along the apical surface of the ventricular zone, cortical progenitors are born from radial glia and undergo a series of symmetric and asymmetric cell divisions to give rise to the cellular diversity throughout the cortex (Dehay and Kennedy 2007). Unlike progenitors of the olfactory bulb that continue to proliferate as they migrate from the dorsal LGE through the rostral migratory stream (Altman and Das 1965; Menezes et al. 1995), interneuronal progenitors born in the MGE appear to complete the last S-phase of their cell cycle prior to beginning their migration into cortical and limbic regions (Polleux et al. 2002; Xu et al. 2003; Xu et al. 2005). Generally speaking, birthdating of GABAergic interneuronal progenitors in rodent and ferret cortex reveals a similar "inside-out" pattern to that established by projection neurons of the same layer. Deeper layer interneurons tend to leave the cell cycle prior to those destined for the superficial layers (Miller 1985; Fairén et al. 1986; Peduzzi 1988). However, this scenario does not appear to hold for the vertically oriented, calretinin-expressing population (Rymar and Sadikot 2007). In addition, when fate is characterized by physiological parameters, there appears to be a time of birthdate-fate dependence within a given layer (Miyoshi et al. 2007).

2.4 Specification of Cortical Interneurons

As similar results are found from MGE transplants directly onto cortical feeder cells, or directly into the cortical plate of neonates in vivo, or homotopic transplants into the MGE in utero, transplantation studies suggest that interneuron subgroup fate is specified based on signaling encountered during their developmental origins (Xu et al. 2004; Butt et al. 2005; Xu et al. 2005). Given the interest in using cortical interneuron transplantations to repress medication intractable seizures (Lindvall and Bjorklund 1992), or even as a drug delivery system (Wichterle et al. 1999), these results are highly encouraging in that subcortical to cortical migration does not appear to be required for most aspects of interneuron fate specification.

Both overlapping and distinct gene expression patterns have been identified between the interneuron-generating regions of the LGE/dCGE and the MGE (Corbin et al. 2003). Nkx2.1, a homeodomain-containing transcription factor, is expressed within the proliferative zone of the MGE and in the more ventrally located preoptic region (Kimura et al. 1996). Nkx2.1 is downregulated in cortical interneuronal progenitors prior to their entry into the cerebral cortex, but is maintained in subsets of striatal interneurons (Marin et al. 2000). Nkx2.1 null mice fail to form a normal MGE, although there is a ventral expansion of the LGE-like tissue (Sussel et al. 1999). At E18.5, Nkx2.1 mutants lack SST and NPY expression in the cortex (Anderson et al. 2001). To determine the requirement of Nkx2.1 for specifying other interneuron subgroups in these perinatal lethal mutants, cortices from E18.5 embryos were dissociated and maintained 2-4 weeks in vitro (Xu et al. 2004). Consistent with studies on the interneuron fate potential of progenitors from the MGE, PV, SST, and NPY were present in cultures of wild-type cortex but absent in those from Nkx2.1 nulls. Transplantation of the MGE-like region of Nkx2.1 nulls onto cortical feeder cultures from normal mice, also failed to give rise to the PV-, SST-, or NPY-expressing subgroups, suggesting that Nkx2.1 is required for the initial specification of these cell types. Interestingly, bipolar CR+ interneurons were plentiful in cortical cultures from Nkx2.1 nulls, consistent with their origin from an Nkx2.1 negative domain.

The requirement of Nkx2.1 for the specification of MGE-derived interneuron subgroups provides a focal point in the search for interneuron fate-specifying factors that act upstream, downstream, and in conjunction with this transcription factor. Initial patterning of Nkx2.1 expression in the ventral telencephalon involves the coordinate actions of the signaling molecules Fgf8 and Sonic Hedgehog (Shh) (Lupo et al. 2006; Storm et al. 2006). Six3 may confer competence of the telencephalic tissue to respond to Shh by inducing Nkx2.1 (Kobayashi et al. 2002), and repression of bone morphogenic protein (BMP) signaling is also required for normal patterning of the Nkx2.1 domain (Anderson et al. 2002a).

Although the patterning role of Shh is largely complete by E11.5 (Kohtz et al. 1998; Fuccillo et al. 2004), some targets of Shh signaling, including Nkx2.1, remain dependent on Shh for their maintenance in MGE progenitors well into the age range of cortical interneuronogenesis (in mouse, roughly E12.5-E16.5). Reductions of Shh signaling in MGE progenitors, essentially all of which normally express Nkx2.1, result in both a large reduction in Nkx2.1 protein detectability despite continued progenitor cycling, and a reduction in the ability of these progenitors to generate PV- or SST-expressing interneurons (Xu et al. 2005). Remarkably, both the Nkx2.1 protein levels, and the generation of SST-expressing interneurons is rescued in telencephalic slices of NestinCre:Shh flox/flox mutants by the restoration of Shh signaling (Xu et al. 2005). This result suggests that interneuron specification remains plastic during the age range of neurogenesis, so that interneuron generation could be altered by a variety of environmental conditions that effect signaling of Shh, FGFs, BMPs in addition to other factors (Yung et al. 2002; Gulacsi and Lillien 2003).

2.4.1 Generation of Interneuron Diversity Within the MGE

While the expression of Nkx2.1 distinguishes the origins of most PV- and SST-expressing interneuron subgroups from that of the vertically oriented CR+ subgroup, less is known about the differential specification of PV- and SST-expressing subgroups within the Nkx2.1 lineage. One possibility, that would be analogous to the neuronal fate determination in the spinal cord (Jessell 2000), is that MGE-derived interneuron subgroups originate from distinct lineages that are separated on the dorso-ventral axis. In fact, Nkx6.2, a transcription factor that contributes to oligodendrocyte generation in the ventral spinal cord (Vallstedt et al. 2005), is selectively expressed in the dorsal-most region of the MGE (Stenman et al. 2003b), and is downregulated in CNS-specific Shh mutant mice that also have a large reduction of PV- and SST-expressing cortical interneurons (Xu et al. 2005). Another study using fate mapping of transgenic mice expressing Cre under both the Nkx2.1 and Nkx6.2 promoters found that approximately 90% of the CB-, SST-, and PV-expressing interneuronal subgroups were labeled (Fogarty et al. 2007). This study also found that the dorsal MGE gives rise to most of the SST+/CR+ colabeled Martinotti interneurons. In support of these findings, transplantation studies found that PV- and SST-expressing interneuron subgroups arise primarily from the ventral and dorsal MGE, respectively (Flames et al. 2007; Wonders et al. 2008). As the expression of Shh signaling effectors Gli1 and Gli2 are enhanced in the dorsal MGE, these results suggest that higher levels of Shh signaling regulate the specification of dorsal MGE cells into SST-expressing subgroups (Wonders et al. 2008). However, as the mRNA expression of Shh would predict a ventral high, dorsal low gradient, and if the above prediction were in fact "true," a mechanism for the enhanced Shh signaling would need to be established.

Like the differential specification of MGE-derived interneuron subgroups, transcriptional regulators that effect interneuron development downstream of Nkx2.1 are beginning to be appreciated. Chief among these genes is Lhx6, a member of the lim-homeodomain family of transcription factors that is strongly expressed in the postmitotic mantle zone of the MGE (Grigoriou et al. 1998). In the spinal cord, lim-homeodomain genes regulate the specification of subgroups of motor neurons (Sharma et al. 1998). Lhx6 expression is undetectable in the telencephalon of Nkx2.1 null mice (Sussel et al. 1999). Lhx6 is expressed in the MGE-derived interneuronal progenitors upon their migration to the cortex (Lavdas et al. 1999; Gong et al. 2003), and its expression is maintained in most PV-expressing and SST-expressing cortical interneurons in adult mice (Fogarty et al. 2007; Liodis et al. 2007). Although there is no difference in the number of GAD67+ cells in the cortex of Lhx6 null mice, these animals exhibit differences in the local, cortical distribution of GAD67+ cells and are completely devoid of PV and SST expression despite the maintained presence of CR+ interneurons (Liodis et al. 2007). Transfection of MGE cells in slice culture with an RNAi construct targeting Lhx6 resulted in a reduction of interneuron migration into the cortex, but no alteration in GABA expression (Alifragis et al. 2004). Although Lhx6 may be dispensable for the expression of GABA in MGE-derived progenitors, it is sufficient for the rescue of PV and SST expression in transplanted, Nkx2.1–/– cells (Du and Anderson unpublished), suggesting that Lhx6 promotes both cortical migration and later aspects of MGE-derived interneuron specification into distinct neurochemical subgroups.

Although the role for continued Lhx6 expression remains unclear, the postnatal expression of the transcription factor Dlx1 is critical for since Dlx1 mutants show a selective loss of some CR-, NPY-, and SST-expressing interneurons beginning around the fourth postnatal week. The transplantation of GFP-expressing MGE progenitors from Dlx1 mutants into wildtype neonatal cortex further showed that this cell loss is due to a cell autonomous requirement for Dlx1 and is preceded by decreased dendritic length and branching. To date, these findings are, may be, the first to describe a transcription factor mutation that cell autonomously alters postnatal development of cortical interneurons (Cobos et al. 2005).

In addition to the issue of interneuron fate determination, slice culture, transgenic mouse, and transplantation experiments have been examining the regulation of interneuron migration from the MGE into the neocortex. First, Slit (ligand) Robo (receptor) interactions appear to drive the cells away from the proliferative zone (Zhu et al. 1999). Second, a combination of chemorepulsion (semaphorin – neuropilin) (Marin et al. 2001), permissive substrate (membrane-bound neuregulin – Erb4), and chemoattractive (diffusible neuregulin – Erb4) signals guide the interneurons into the cortex (Flames et al. 2004). As interneurons reach the cortex, they tend to parse into three streams that run above and below the cortical plate and in the deep intermediate zone. They then turn from their predominantly tangential orientations to migrate radially into the cortical plate (Metin et al. 2006). It remains to be determined whether a cellular substrate such as radial glia, or the axonal or dendritic processes of pyramidal neurons, are mediating this migration. Recent evidence suggests that chemokine signaling via Cxcl12/Cxcr4 signaling initially prevents the interneuron invasion into the cortical plate (Lopez-Bendito et al. in press; Li et al. in press). However, it remains unclear whether interneuron subtypes are following subtype-specific cues to determine the precise location of their terminal differentiation. Alternatively, the migrating interneurons may use maturational/timing-dependent cues that allow competency to enter the cortical plate, then like-subtype chemorepulsion to distribute themselves evenly across a given cortical region and layer.

So where do we stand in the process of understanding the molecular regulation of cortical interneuron fate determination? The temporal and spatial origins of cortical interneurons and their migratory pathways are fairly well described, particularly in rodents. The differential spatial origins within the subpallium correlate with differences in the expression of a few fate-altering proteins that result in the specification of neurochemically and physiologically distinct interneuron subgroups.

Moving forward, these findings are being extended by methods that permit the systematic study of interneuron fate determination. For example, neurochemical aspects of interneuron subgroup fate determination are maintained in vitro when interneuron progenitors are plated over a feeder layer of cortical cells. This technique provides a relatively high throughput way to study the molecular regulation of subgroup fate determination and cortical influences on interneuron differentiation (Xu et al. 2004; Xu et al. 2005). Of particular interest is whether aspects of interneuron subgroup physiology and connectivity can be meaningfully studied using the interneuron progenitor-cortical culture method.

More importantly, techniques have been developed to study interneuron fate determination in vivo. The most elegant method presently available is that of transplanting genetically labeled interneuron progenitors homotopically by in utero transplantation (Wichterle et al. 2001; Butt et al. 2005). This methodology can also be performed with genetically altered progenitors which, in addition to interneuron transplants directly into the cortical plate (Cobos et al. 2005; Wonders et al. 2008), will provide critical data on the cell autonomous regulation of interneuron development. In utero electroporation of marker genes (Borrell et al. 2005), a technique likely to be extended to gain and loss of function studies, provides yet another tool for the in vivo examination of embryonic manipulations on fate determination. Meanwhile, genetic differences within mature cortical interneuron populations are beginning to be elucidated (Sugino et al. 2006), enhancing the ability to perform specific labeling of subgroups of cortical interneurons in the adult (Oliva et al. 2000; Meyer et al. 2002; Ma et al. 2006). In addition, with the advent of directed differentiation of embryonic stem cells toward neural progenitors, it will be possible to study the transcriptional alterations and epigenetic modifications that occur as a cell goes from a pluripotent state to a specific interneuron with distinct neurochemistry, morphology, and electrophysiological properties. In sum, the field is poised to bridge the gap between the molecular control of interneuron fate determination and the molecular basis of interneuron connectivity and physiology.

References

Alifragis P, Liapi A, Parnavelas JG (2004) Lhx6 regulates the migration of cortical interneurons from the ventral telencephalon but does not specify their GABA phenotype. J Neurosci 24:5643–5648

Altman J, Das GD (1965) Autoradiographic and histological evidence of postnatal hippocampal neurogenesis in rats. J Comp Neurol 124:319–335

Anderson SA, Eisenstat DD, Shi L, Rubenstein JL (1997) Interneuron migration from basal forebrain to neocortex: dependence on Dlx genes. Science 278:474–476

Anderson S, Mione M, Yun K, Rubenstein JLR (1999) Differential origins of projection and local circuit neurons: role of Dlx genes in neocortical interneuronogenesis. Cereb Cortex 9:646–654

Anderson SA, Marin O, Horn C, Jennings K, Rubenstein JL (2001) Distinct cortical migrations from the medial and lateral ganglionic eminences. Development 128:353–363

Anderson RM, Lawrence AR, Stottmann RW, Bachiller D, Klingensmith J (2002a) Chordin and noggin promote organizing centers of forebrain development in the mouse. Development 129:4975–4987

Anderson SA, Kaznowski CE, Horn C, Rubenstein JL, McConnell SK (2002b) Distinct origins of neocortical projection neurons and interneurons in vivo. Cereb Cortex 12:702–709

Ang ES Jr, Haydar TF, Gluncic V, Rakic P (2003) Four-dimensional migratory coordinates of GABAergic interneurons in the developing mouse cortex. J Neurosci 23:5805–5815

Bellion A, Wassef M, Metin C (2003) Early differences in axonal outgrowth, cell migration and GABAergic differentiation properties between the dorsal and lateral cortex. Cereb Cortex 13:203–214

Borrell V, Yoshimura Y, Callaway EM (2005) Targeted gene delivery to telencephalic inhibitory neurons by directional in utero electroporation. J Neurosci Methods 143:151–158

Butt SJ, Fuccillo M, Nery S, Noctor S, Kriegstein A, Corbin JG, Fishell G (2005) The temporal and spatial origins of cortical interneurons predict their physiological subtype. Neuron 48:591–604

Cobos I, Calcagnotto ME, Vilaythong AJ, Thwin MT, Noebels JL, Baraban SC, Rubenstein JL (2005) Mice lacking Dlx1 show subtype-specific loss of interneurons, reduced inhibition and epilepsy. Nat Neurosci 8:1059–1068

Corbin JG, Rutlin M, Gaiano N, Fishell G (2003) Combinatorial function of the homeodomain proteins Nkx2.1 and Gsh2 in ventral telencephalic patterning. Development 130:4895–4906

de Carlos JA, López-Mascaraque L, Valverde F (1996) Dynamics of cell migration from the lateral ganglionic eminence in the rat. J Neurosci 16:6146–6156

DeDiego I, Smith-Fernandez A, Fairen A (1994) Cortical cells that migrate beyond area boundaries: characterization of an early neuronal population in the lower intermediate zone of prenatal rats. Eur J NeuroSci 6:983–997

DeFelipe J (1997) Types of neurons, synaptic connections and chemical characteristics of cells immunoreactive for calbindin-D28K, parvalbumin and calretinin in the neocortex. J Chem Neuroanat 14:1–19

DeFelipe J, Jones EJ (eds) (1988) Cajal on the cerebral cortex. Oxford University Press, New York

Dehay C, Kennedy H (2007) Cell-cycle control and cortical development. Nat Rev Neurosci 8:438–450

Fairén A, Cobas A, Fonseca M (1986) Times of generation of glutamic acid decarboxylase immunoreactive neurons in mouse somatosensory cortex. J Comp Neurol 251:67–83

Flames N, Long JE, Garratt AN, Fischer TM, Gassmann M, Birchmeier C, Lai C, Rubenstein JL, Marin O (2004) Short- and long-range attraction of cortical GABAergic interneurons by neuregulin-1. Neuron 44:251–261

Flames N, Pla R, Gelman DM, Rubenstein JL, Puelles L, Marin O (2007) Delineation of multiple subpallial progenitor domains by the combinatorial expression of transcriptional codes. J Neurosci 27:9682–9695

Fogarty M, Grist M, Gelman D, Marin O, Pachnis V, Kessaris N (2007) Spatial genetic patterning of the embryonic neuroepithelium generates GABAergic interneuron diversity in the adult cortex. J Neurosci 27:10935–10946

Fuccillo M, Rallu M, McMahon AP, Fishell G (2004) Temporal requirement for hedgehog signaling in ventral telencephalic patterning. Development 131:5031–5040

Gonchar Y, Burkhalter A (1997) Three distinct families of GABAergic neurons in rat visual cortex. Cereb Cortex 7:347–358

Gong S, Zheng C, Doughty ML, Losos K, Didkovsky N, Schambra UB, Nowak NJ, Joyner A, Leblanc G, Hatten ME, Heintz N (2003) A gene expression atlas of the central nervous system based on bacterial artificial chromosomes. Nature 425:917–925

Götz M, Williams BP, Bolz J, Price J (1995) The specification of neuronal fate: a common precursor for neurotransmitter subtypes in the rat cerebral cortex in vitro. Eur J NeuroSci 7:889–898

Grigoriou M, Tucker AS, Sharpe PT, Pachnis V (1998) Expression and regulation of Lhx6 and Lhx7, a novel subfamily of LIM homeodomain encoding genes, suggests a role in mammalian head development. Development 125:2063–2074

Gulacsi A, Lillien L (2003) Sonic hedgehog and bone morphogenetic protein regulate interneuron development from dorsal telencephalic progenitors in vitro. J Neurosci 23:9862–9872

He W, Ingraham C, Rising L, Goderie S, Temple S (2001) Multipotent stem cells from the mouse basal forebrain contribute GABAergic neurons and oligodendrocytes to the cerebral cortex during embryogenesis. J Neurosci 21:8854–8862

Hensch TK (2005) Critical period plasticity in local cortical circuits. Nat Rev Neurosci 6:877–888

Hevner RF, Daza RA, Englund C, Kohtz J, Fink A (2004) Postnatal shifts of interneuron position in the neocortex of normal and reeler mice: evidence for inward radial migration. Neuroscience 124:605–618

Jessell TM (2000) Neuronal specification in the spinal cord: inductive signals and transcriptional codes. Nat Rev Genet 1:20–29

Jimenez D, Lopez-Mascaraque LM, Valverde F, De Carlos JA (2002) Tangential migration in neocortical development. Dev Biol 244:155–169

Kawaguchi Y, Kubota Y (1997) GABAergic cell subtypes and their synaptic connections in rat frontal cortex. Cereb Cortex 7:476–486

Kessaris N, Fogarty M, Iannarelli P, Grist M, Wegner M, Richardson WD (2006) Competing waves of oligodendrocytes in the forebrain and postnatal elimination of an embryonic lineage. Nat Neurosci 9:173–179

Kimura S, Hara Y, Pineau T, Fernandez-Salguero P, Fox CH, Ward JM, Gonzalez FJ (1996) The T/ebp null mouse: thyroid-specific enhancer-binding protein is essential for the organogenesis of the thyroid, lung, ventral forebrain, and pituitary. Genes Dev 10:60–69

Kobayashi D, Kobayashi M, Matsumoto K, Ogura T, Nakafuku M, Shimamura K (2002) Early subdivisions in the neural plate define distinct competence for inductive signals. Development 129:83–93

Kohtz JD, Baker DP, Corte G, Fishell G (1998) Regionalization within the mammalian telencephalon is mediated by changes in responsiveness to Sonic Hedgehog. Development 125:5079–5089

Lavdas AA, Grigoriou M, Pachnis V, Parnavelas JG (1999) The medial ganglionic eminence gives rise to a population of early neurons in the developing cerebral cortex. J Neurosci 19:7881–7888

Letinic K, Zoncu R, Rakic P (2002) Origin of GABAergic neurons in the human neocortex. Nature 417:645–649

Lindvall O, Bjorklund A (1992) Intracerebral grafting of inhibitory neurons. A new strategy for seizure suppression in the central nervous system. Adv Neurol 57:561–569

Liodis P, Denaxa M, Grigoriou M, Akufo-Addo C, Yanagawa Y, Pachnis V (2007) Lhx6 activity is required for the normal migration and specification of cortical interneuron subtypes. J Neurosci 27:3078–3089

Lopez-Bendito G, Sturgess K, Erdelyi F, Szabo G, Molnar Z, Paulsen O (2004) Preferential origin and layer destination of GAD65-GFP cortical interneurons. Cereb Cortex 14:1122–1133

Lupo G, Harris WA, Lewis KE (2006) Mechanisms of ventral patterning in the vertebrate nervous system. Nat Rev Neurosci 7:103–114

Ma Y, Hu H, Berrebi AS, Mathers PH, Agmon A (2006) Distinct subtypes of somatostatin-containing neocortical interneurons revealed in transgenic mice. J Neurosci 26:5069–5082

Marin O, Rubenstein JL (2001) A long, remarkable journey: tangential migration in the telencephalon. Nature Reviews Neuroscience 2:780–790

Marin O, Anderson SA, Rubenstein JL (2000) Origin and molecular specification of striatal interneurons. J Neurosci 20:6063–6076

Marin O, Yaron A, Bagri A, Tessier-Lavigne M, Rubenstein JL (2001) Sorting of striatal and cortical interneurons regulated by semaphorin-neuropilin interactions. Science 293:872–875

Menezes JR, Smith CM, Nelson KC, Luskin MB (1995) The division of neuronal progenitor cells during migration in the neonatal mammalian forebrain. Mol Cell Neurosci 6:496–508

Metin C, Baudoin JP, Rakic S, Parnavelas JG (2006) Cell and molecular mechanisms involved in the migration of cortical interneurons. Eur J NeuroSci 23:894–900

Meyer G, Soria JM, Martínez-Galán JR, Martín-Clemente B, Fairén A (1998) Different origins and developmental histories of transient neurons in the marginal zone of the fetal and neonatal rat cortex. J Comp Neurol 397:493–518

Meyer AH, Katona I, Blatow M, Rozov A, Monyer H (2002) In vivo labeling of parvalbumin-positive interneurons and analysis of electrical coupling in identified neurons. J Neurosci 22:7055–7064

Miller MW (1985) Cogeneration of retrogradely labeled corticocortical projection and GABA-immunoreactive local circuit neurons in cerebral cortex. Brain Res 355:187–192

Miyoshi G, Butt SJ, Takebayashi H, Fishell G (2007) Physiologically distinct temporal cohorts of cortical interneurons arise from telencephalic Olig2-expressing precursors. J Neurosci 27:7786–7798

Monyer H, Markram H (2004) Interneuron Diversity series: Molecular and genetic tools to study GABAergic interneuron diversity and function. Trends Neurosci 27:90–97

Nery S, Fishell G, Corbin JG (2002) The caudal ganglionic eminence is a source of distinct cortical and subcortical cell populations. Nat Neurosci 5:1279–1287

Nery S, Corbin JG, Fishell G (2003) Dlx2 progenitor migration in wild type and nkx2.1 mutant telencephalon. Cereb Cortex 13:895–903

Oliva AA Jr, Jiang M, Lam T, Smith KL, Swann JW (2000) Novel hippocampal interneuronal subtypes identified using transgenic mice that express green fluorescent protein in GABAergic interneurons. J Neurosci 20:3354–3368

Owens DF, Kriegstein AR (2002) Is there more to GABA than synaptic inhibition? Nat Rev Neurosci 3:715–727

Parnavelas JG (2000) The origin and migration of cortical neurones: new vistas. Trends Neurosci 23:126–131

Peduzzi JD (1988) Genesis of GABA-immunoreactive neurons in the ferret visual cortex. J Neurosci 8:920–931

Polleux F, Whitford KL, Dijkhuizen PA, Vitalis T, Ghosh A (2002) Control of cortical interneuron migration by neurotrophins and PI3-kinase signaling. Development 129:3147–3160

Porteus MH, Bulfone A, Liu JK, Lo LC, Rubenstein JLR (1994) DLX-2, MASH-1, and MAP-2 expression and bromodeoxyuridine incorporation define molecularly distinct cell populations in the embryonic mouse forebrain. J Neuroscience 44:6370–6383

Rakic S, Zecevic N (2003) Early oligodendrocyte progenitor cells in the human fetal telencephalon. Glia 41:117–127

Rogers JH (1992) Immunohistochemical markers in rat cortex: co-localization of calretinin and calbindin-D28k with neuropeptides and GABA. Brain Res 587:147–157

Rymar VV, Sadikot AF (2007) Laminar fate of cortical GABAergic interneurons is dependent on both birthdate and phenotype. J Comp Neurol 501:369–380

Sharma K, Sheng HZ, Lettieri K, Li H, Karavanov A, Potter S, Westphal H, Pfaff SL (1998) LIM homeodomain factors Lhx3 and Lhx4 assign subtype identities for motor neurons. Cell 95:817–828

Stenman J, Toresson H, Campbell K (2003a) Identification of two distinct progenitor populations in the lateral ganglionic eminence: implications for striatal and olfactory bulb neurogenesis. J Neurosci 23:167–174

Stenman JM, Wang B, Campbell K (2003b) Tlx controls proliferation and patterning of lateral telencephalic progenitor domains. J Neurosci 23:10568–10576

Storm EE, Garel S, Borello U, Hebert JM, Martinez S, McConnell SK, Martin GR, Rubenstein JL (2006) Dose-dependent functions of Fgf8 in regulating telencephalic patterning centers. Development 133:1831–1844

Sugino K, Hempel CM, Miller MN, Hattox AM, Shapiro P, Wu C, Huang ZJ, Nelson SB (2006) Molecular taxonomy of major neuronal classes in the adult mouse forebrain. Nat Neurosci 9:99–107

Sussel L, Marin O, Kimura S, Rubenstein JL (1999) Loss of Nkx2.1 homeobox gene function results in a ventral to dorsal molecular respecification within the basal telencephalon: evidence for a transformation of the pallidum into the striatum. Development 126:3359–3370

Taglialatela P, Soria JM, Caironi V, Moiana A, Bertuzzi S (2004) Compromised generation of GABAergic interneurons in the brains of Vax1-/- mice. Development 131:4239–4249

Tamamaki N, Fujimori KE, Takauji R (1997) Origin and route of tangentially migrating neurons in the developing neocortical intermediate zone. J Neurosci 17:8313–8323

Valcanis H, Tan SS (2003) Layer Specification of Transplanted Interneurons in Developing Mouse Neocortex. J Neurosci 23:5113–5122

Vallstedt A, Klos JM, Ericson J (2005) Multiple dorsoventral origins of oligodendrocyte generation in the spinal cord and hindbrain. Neuron 45:55–67

Wang XJ, Tegner J, Constantinidis C, Goldman-Rakic PS (2004) Division of labor among distinct subtypes of inhibitory neurons in a cortical microcircuit of working memory. Proc Natl Acad Sci U S A 101:1368–1373

Whittington MA, Traub RD (2003) Interneuron diversity series: inhibitory interneurons and network oscillations in vitro. Trends Neurosci 26:676–682

Wichterle H, Garcia-Verdugo JM, Herrera DG, Alvarez-Buylla A (1999) Young neurons from medial ganglionic eminence disperse in adult and embryonic brain. Nat Neurosci 2:461–466

Wichterle H, Turnbull DH, Nery S, Fishell G, Alvarez-Buylla A (2001) In utero fate mapping reveals distinct migratory pathways and fates of neurons born in the mammalian basal forebrain. Development 128:3759–3771

Wonders CP, Anderson SA (2006) The origin and specification of cortical interneurons. Nat Rev Neurosci 7; 687–696

Wonders CP, Taylor L, Welagen J, Mbata IC, Xiang JZ, Anderson SA (2008) A spatial bias for the origins of interneuron subgroups within the medial ganglionic eminence. Dev Biol 314:127–136

Xu Q, De La Cruz E, Anderson SA (2003) Cortical interneuron fate determination: diverse sources for distinct subtypes? Cereb Cortex 13:670–676

Xu Q, Cobos I, De La Cruz E, Rubenstein JL, Anderson SA (2004) Origins of cortical interneuron subtypes. J Neurosci 24:2612–2622

Xu Q, Wonders CP, Anderson SA (2005) Sonic hedgehog maintains the identity of cortical interneuron progenitors in the ventral telencephalon. Development 132:4987–4998

Xu X, Roby KD, Callaway EM (2006) Mouse cortical inhibitory neuron type that coexpresses somatostatin and calretinin. J Comp Neurol 499:144–160

Xu Q, Tam M, Anderson SA (2008) Fate mapping Nkx2.1-lineage cells in the mouse telencephalon. J Comp Neurol 506:16–29

Yozu M, Tabata H, Nakajima K (2005) The caudal migratory stream: a novel migratory stream of interneurons derived from the caudal ganglionic eminence in the developing mouse forebrain. J Neurosci 25:7268–7277

Yung SY, Gokhan S, Jurcsak J, Molero AE, Abrajano JJ, Mehler MF (2002) Differential modulation of BMP signaling promotes the elaboration of cerebral cortical GABAergic neurons or oligodendrocytes from a common sonic hedgehog-responsive ventral forebrain progenitor species. Proc Natl Acad Sci U S A 99:16273–16278

Zecevic N, Rakic P (2001) Development of layer I neurons in the primate cerebral cortex. J Neurosci 21:5607–5619

Zhu Y, Li H, Zhou L, Wu JY, Rao Y (1999) Cellular and molecular guidance of GABAergic neuronal migration from an extracortical origin to the neocortex. Neuron 23:473–485

Chapter 3
Role of Spontaneous Activity in the Maturation of GABAergic Synapses in Embryonic Spinal Circuits

Carlos E. Gonzalez-Islas and Peter Wenner

It has become increasingly evident that neural activity is indispensable for synapse and network formation in many parts of the central nervous system. During development, GABAergic neurotransmission contributes greatly to the embryonic neural hyper-excitability due to its depolarizing nature at this stage. However, the precise way in which embryonic neural activity shapes GABAergic transmission is just beginning to be unveiled.

Gamma-aminobutyric acid (GABA) is the main inhibitory transmitter in the adult brain; however in early development, it is actually depolarizing and excitatory. Therefore, glutamatergic and GABAergic ionotropic transmission excites postsynaptic cells, making these recurrently connected networks very excitable. This transient developmental condition leads to an almost epileptic-like activity known as spontaneous network activity (SNA) that is experienced in most, if not all, developing circuits. In this chapter, we will discuss the depolarizing nature of GABA, how it is critical to the generation of SNA in the spinal cord, and how GABA and SNA interact and influence each other. Finally, we will discuss how SNA may drive the maturation of GABAergic synaptic strength through a process known as homeostatic synaptic plasticity. This process is also important in the maturation of glutamatergic synaptic strength in a manner that coordinates the development of excitatory and soon to be inhibitory systems. This process is likely to be important in establishing a balance between the two systems that will be critical for the appropriate behavior of mature networks.

Fast GABAergic action in the mature CNS is hyperpolarizing and inhibitory in most cases; it can be considered as a necessary break for networks driven primarily by the excitatory action of glutamate. However, the excitatory actions of GABAergic transmission during the establishment and early maturation of the neuronal circuits are critical for the many roles of this transmitter in development (Singer and Berger 2000; Ben-Ari et al. 2007; Akerman and Cline 2007). It is now appreciated that

---

C.E. Gonzalez-Islas and P. Wenner (✉)
Department of Physiology, School of Medicine, Emory University, 615 Michael St, Atlanta, GA, 30322, USA
e-mail: pwenner@emory.edu

GABA signaling, likely due to its early depolarizing nature, is involved in directing multiple developmental processes, including cell proliferation, migration, and differentiation; establishment of synaptic connections and their refinement; and possibly in the depolarizing to hyperpolarizing conversion of the $GABA_A$ response itself (Ben-Ari et al. 2007; Owens and Kriegstein 2002; Ge et al. 2006; Akerman and Cline 2007; Kandler and Gillespie 2005).

Many of the specialized tasks, executed by GABAergic transmission in the course of embryonic development, are due to its ability to depolarize the membrane potential in embryonic cells. These fast GABAergic inward currents are mediated by the activation of $GABA_A$ receptors ($GABA_A$-Rs). GABA released from presynaptic vesicles activates $GABA_A$-Rs, thereby opening conductances to chloride (and to a lesser extent bicarbonate ions – HCO_3^-). In adults, intracellular chloride concentration is low; thus, GABAergic transmission allows chloride to rush into the cell, thereby hyperpolarizing it. In early development, however, intracellular chloride concentration is higher than in the adult, making the reversal potential for chloride more depolarized than the resting membrane potential, resulting in chloride efflux and depolarization (Payne et al. 2003; Ben-Ari et al. 2007). Intracellular chloride accumulations in the young cells are thought to occur due to the stronger expression of an Na^+-K^+-Cl^- cotransporter (NKCC1) in early development. NKCC1 uses the energy stored in concentration gradients across the cell membrane to transport the various ions. Later in development, intracellular chloride concentrations drop as a result of the reduced function of NKCC1 and the concurrent increase in function of the K^+–Cl^- cotransporter (KCC2), the main chloride carrier in mature cells. KCC2, in contrast to NKCC1, extrudes chloride using energy from the potassium ion concentration gradient, and thereby reduces intracellular chloride concentration. Owing to the transient capacity of chloride currents to depolarize neurons during a limited developmental window, GABA is in a strong position to regulate activity-dependent processes during the crucial period when neural differentiation and circuit formation are occurring in the CNS.

As neuronal networks are recurrently connected, and because glutamate and GABA are both largely excitatory during early development, when one set of neurons becomes active this tends to spread and recruit the rest of the network. These developing circuits are therefore highly excitable and produce spontaneous bursts of network activity present in most if not all developing networks, including the spinal cord, hippocampus, and retina (O'Donovan 1999; Feller 1999; Ben-Ari et al. 2007). In the embryonic spinal cord, this spontaneous network activity (SNA) acts to recruit the majority of spinal neurons into bouts or episodes of activity (Fig. 3.1). Because motoneurons innervating limb muscles are also recruited during the episodes, SNA drives spontaneous limb movements such as those observed in the human embryo and fetus. Embryonic movements can be found in virtually all vertebrate species, demonstrating the importance of such activity. SNA and the movements generated by it are important for the development of limb muscle and joints (Hall and Herring 1990; Jarvis et al. 1996; Persson 1983; Roufa and Martonosi 1981) and are involved in motoneuron axon guidance (Hanson and Landmesser 2004; Hanson et al. 2008). Although early studies blocking SNA did not show clear effects on

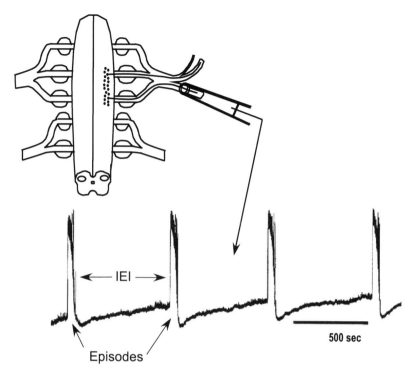

Fig. 3.1 Schematic of muscle nerve recordings from an isolated chick embryo spinal cord trace showing regularly occurring episodes of SNA separated by inter-episode intervals (IEI)

circuit behavior, we have recently shown that this activity is important in setting the strength of the synaptic connections for both excitatory and inhibitory inputs in the developing spinal cord (Gonzalez-Islas and Wenner 2006).

Embryonic limb movements in different species are restricted to a particular interval during development and are organized in short episodes of motility followed by periods of relative calm (Hamburger 1977; Bekoff et al. 1975). Similar activity patterns can be observed in isolated spinal cord preparations, in which many spinal neurons become active during episodes that last approximately 60 s (Fig. 3.1, O'Donovan et al. 1998). These episodes are followed by longer periods of quiescence called inter-episode intervals (IEI, minutes), in which the spinal neurons are relatively silent. Episodes of SNA can first be detected in the chick embryo at about embryonic day 4 or 5 (E4-5) as very simple spontaneous episodic events, consisting of single recurring depolarizing events, evolving progressively into more recurrent multicycle episodes as the network matures (Milner and Landmesser 1999; O'Donovan 1987). In general, spontaneous neuronal activity is a characteristic feature of developing neuronal systems; however, the type of activity manifested depends on the level of differentiation of the individual neurons and on the degree to which these neurons constitute themselves into networks.

For example, before the establishment of chemical synaptic connections, spontaneous spiking can be seen in isolated neurons, and coordinated calcium transients can be recorded in groups of electrically coupled neurons (Yuste et al. 1995; Gu and Spitzer 1997). Later in development, as soon as chemical synapses are established, a different kind of activity appears, now driven by the network (SNA), which mainly depends on chemical synaptic transmission, but not on detailed or specific connectivity, nor on pacemaker cells (O'Donovan 1999).

Immediately following an episode of SNA, the network is depressed, but it slowly recovers in the IEI (Fig. 3.2). Several observations by the O'Donovan lab are consistent with this idea. If reflexes are stimulated early in the IEI, they are weak, but then they strengthen later in the interval, once the network recovers from the network-induced depression (Fedirchuk et al. 1999; Chub and O'Donovan 2001; Gonzalez-Islas and Wenner 2006) (Fig. 3.2a). Correspondingly, these stimulations are more capable of evoking episodes later in the interval. Similarly, it is known that quantal amplitude is depressed immediately after an episode and progressively recovers through the course of the IEI (Ritter et al. 1999; Chub and O'Donovan 2001; Gonzalez-Islas and Wenner 2006). Furthermore, at the beginning of the IEI, spinal neurons are hyperpolarized by ~10 mV and progressively repolarize following a depolarizing ramp of 0.5–1 mV/min (Fig. 3.2b, Chub and O'Donovan 2001). All of these observations can be explained as the progressive increase in the

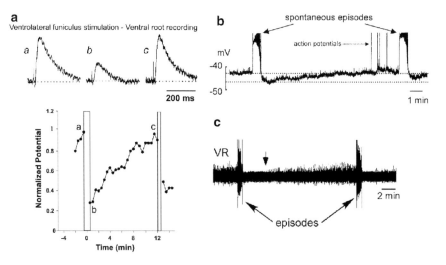

Fig. 3.2 Modulation of excitability in the inter-episode interval. (**a**) Response from ventral root recording (motoneuron population) following stimulation of the ventrolateral funiculus every 30 s. Strong responses are observed just before episodes (a, c), but depressed responses occur right after an episode (b) (Fedirchuk et al. 1999). (**b**) Whole cell recording from a spinal neuron showing a depolarizing ramp potential developing in the IEI and leading to spikes at the end of the IEI (Chub and O'Donovan 2001). (**c**) High-pass filtered ventral root (VR) recording showing the progressive development of motoneuron spiking activity in the IEI (Wenner and O'Donovan 2001)

functional connectivity and in the excitability of the spinal network. After an episode in which spinal neurons are relatively hyperpolarized and synaptic strength is at its weakest, there is virtually no spiking activity. As the IEI progresses and neurons become more depolarized, some spinal neurons begin to reach threshold and fire action potentials (Fig. 3.2c), (Wenner and O'Donovan 2001). As neurons become more depolarized and synaptic strength increases, the motoneuron discharge becomes more vigorous, and a specific class of interneurons that receive direct input from the recurrent collaterals of motoneurons are recruited, which then trigger full-blown episodes (Wenner and O'Donovan 2001).

What produces the modulation of excitability in the IEI? GABAergic synaptic transmission occupies a central role in this modulation of network excitability. The source of the ramp depolarization described above involves a GABAergic current that strengthens progressively in the IEI (Chub and O'Donovan 1998). Further, strong modulations of both GABAergic-evoked potentials and GABAergic miniature postsynaptic currents (mPSCs) are observed in the IEI (Tabak et al. 2001; Gonzalez-Islas and Wenner 2006). The modulation of GABAergic currents therefore significantly contributes to the progressive depolarization and increased synaptic strength that occurs in the intervals between episodes of SNA. The modulation of GABAergic currents comes about, at least in part, as a result of the activity of the developmentally regulated chloride transporter NKCC1 (Fig. 3.3). Intracellular chloride undergoes significant changes during SNA and the IEI. During SNA, $GABA_A$ receptors are

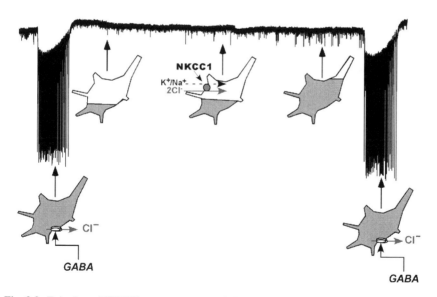

Fig. 3.3 Episode and NKCC1 cotransporter modulation of intracellular chloride concentration. Trace of motoneuron whole cell voltage clamp recording showing mPSCs getting larger in the IEI with schematics of chloride efflux during episodes and chloride re-accumulation by NKCC1 during the IEIs

activated, opening significant chloride conductances. These episodes can last over a minute, and allow significant efflux of chloride ions, so much so that intracellular chloride concentration is reduced during the episode by about 15 mM (Chub and O'Donovan 2001). This translates to a less depolarized chloride equilibrium potential and therefore to a reduction in the driving force for chloride efflux. This then accounts for the reduction in GABAergic currents observed right after an episode. In fact, the reduction in driving force that occurs during the episode may contribute to the episode's termination. Following an episode of SNA, intracellular chloride concentration starts to rise because of an influx of chloride, driven by the activity of NKCC1 (Chub et al. 2006; Marchetti et al. 2005). As intracellular chloride concentration rises, the driving force for this ion also increases, resulting in an increase in GABAergic current strength. As these currents get stronger, enough motoneurons reach threshold to recruit GABAergic R-interneurons and generate an episode through a stochastic process (Tabak et al. 2000, Marchetti et al. 2005). The importance of $GABA_A$ transmission for SNA can be observed by blocking GABAergic currents at E10; the frequency of the episodes becomes slower and highly variable (Chub and O'Donovan 1998). Alternatively, blockade of glutamatergic and cholinergic antagonists at this stage initially slows SNA frequency, but then SNA recovers to near the predrug level of activity and shows little variability. We assert that GABAergic circuits sustain a consistent periodicity because the progressive increase in GABAergic currents strengthens GABAergic synapses and depolarizes the spinal neurons, which progressively increases the likelihood of triggering the next episode.

As described above, GABAergic currents influence the periodicity of the episodes of SNA through changes in intracellular chloride. Turned around, one could say that SNA regulates GABAergic synaptic strength, weakening it after an episode and allowing it to recover in the minutes between episodes. Is it possible that SNA regulates GABAergic synaptic strength over a longer period of time; in other words, could the network assess SNA levels over days and adjust GABAergic synaptic strength to compensate for any perturbations from a set level of activity? It has long been known that activity is an important factor in the process of formation and modification of neuronal circuits (Bliss and Lømo 1973; see Katz and Shatz 1996; Malenka and Nicoll 1999, for review). Currently, a number of studies have focused on how such activity modifies synaptic strength and what the participating mechanisms are. The role of activity as a building block of the CNS has been a central focus of neuroscience. Does activity have a permissive role, refining structures built by a predetermined plan, or does activity do more? In other words, how much experience is important in the formation of the nervous system? The first breakthrough in this debate came from the discovery of long-lasting synaptic modifications, such as long-term potentiation or long-term depression, that are usually synapse-specific and depend on correlation between pre- and postsynaptic firing (Abbott and Nelson, 2000). Correlation-based [Hebbian] rules for the use-dependent modification of synaptic strength have been very enlightening; they constitute the best model of how information is stored in the nervous system (Stent 1973; Hawkins et al. 1993; Malenka and Nicoll, 1993; Linden and Connor 1995) and underlie the refinement of neural connections during development (Shatz 1990; Miller 1994; Cline 1991; Yao and Dan 2005). The problem with this kind of rule lies in its positive feedback nature, as effective synapses are strengthened and less effective ones weakened, the former should continually become more effective and the latter tend to disappear. This is likely to destabilize postsynaptic firing rates since it increases them excessively.

One way to overcome this problem would be the presence of a mechanism that ensures that the cell or network remains within a range of activity that is physiologically appropriate, homeostatically maintaining this activity level. The process of maintaining activity within a certain range has been termed "homeostatic plasticity" and the underlying mechanisms include compensatory regulation of neuronal excitability (LeMasson et al. 1993; Turrigiano et al. 1994; Marder and Goaillard 2006), and synaptic strength (Burrone and Murthy 2003; Rich and Wenner 2006; Turrigiano 2007; Davis 2006). Many studies have focused on compensatory changes in synaptic strength. In cultured neuronal networks, when activity levels were either reduced or increased for two days, compensatory changes were observed in the strength of both AMPAergic and GABAergic synapses (Turrigiano and Nelson 2004). Consequently, when activity levels were reduced, AMPAergic synaptic strength increased and GABAergic synaptic strength decreased (Kilman et al. 2002; O'Brien et al. 1998; Turrigiano et al. 1998). On the contrary, when activity levels were increased, AMPA synaptic strength decreased (Lissin et al 1998; O'Brien et al. 1998; Turrigiano et al. 1998). In each case, the change in synaptic strength acted in a direction that tended to restore the original activity levels. In this way, the network activity could be regulating itself homeostatically through changes in the synaptic strength of the inputs onto the cells in the network. Further, in many cases, all of the excitatory inputs onto a neuron appeared to be increased after activity reduction. Such enhancement of synaptic strength was observed across the entire distribution of mPSCs and was therefore designated as synaptic scaling (Turrigiano et al. 1998).

During development, neurons grow in size, synaptic connections are added or removed, and synaptic strengths change. Any of these transformations could perturb network activity levels in the circuits in which the cells reside. Therefore, we hypothesized that in the formation of neural circuits, homeostatic mechanisms will be at play, and they will maintain the existing activity of these developing networks, i.e., spontaneous network activity. We consequently tested the possibility that embryonic SNA regulated synaptic strength in a homeostatic manner (Gonzalez-Islas and Wenner 2006). We reduced SNA by injecting the sodium channel blocker lidocaine *in ovo* (chick embryo). After chronically reducing SNA for two days (E8-10), we found that the synaptic strength of AMPAergic inputs was increased in a compensatory manner; however, no changes in passive membrane properties were observed. AMPAergic mPSC amplitude and frequency increased following the two-day reduction of SNA. In addition, GABAergic mPSC amplitude increased, and synaptic scaling was observed (Fig. 3.4). Although opposite to the finding in culture, in which IPSCs are hyperpolarizing, in the developing system, this is a compensatory change because GABA is depolarizing and excitatory in embryonic

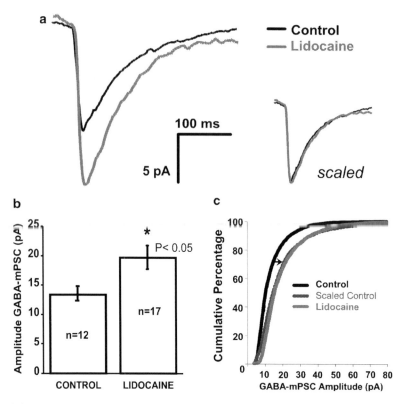

Fig. 3.4 GABAergic mPSC amplitude increase following activity reductions *in ovo*. (**a**) Average $GABA_A$ mPSC in control and lidocaine-treated embryos. The mPSC kinetics are unchanged, but the amplitude increases in treated embryos. (**b**) Bar chart shows the increase in GABAergic mPSC amplitude. (**c**) Cumulative distribution of GABAergic mPSC amplitudes in control and treated embryos can be scaled to match each other using a multiplicative function. Adapted from Gonzalez-Islas and Wenner 2006

spinal neurons. Therefore, if the function of the process is to recover activity levels through changes in synaptic strength, then the network should strengthen both types of excitatory input (GABA and glutamate). These findings suggest that a reduction in network activity can regulate the synaptic strength of mPSCs in a compensatory direction.

What are the mechanisms that underlie the increases in AMPA and GABA quantal amplitude following activity perturbations? In cultured cells, changes have been described for both postsynaptic receptors and transmitter filling of presynaptic vesicles (Rich and Wenner 2007; Turrigiano 2007). AMPA receptors increase and GABA receptors decrease in number, and correspondingly more glutamate per vesicle has been reported. It is not clear whether these mechanisms will be at play in the lidocaine-treated chick embryos.

Interestingly, we have evidence that supports a completely novel mechanism underlying the increase in GABA quantal amplitude. As mentioned above, it is known that in embryonic spinal neurons, the reversal potential for chloride is modulated by an episode of SNA and over minutes in the interval between episodes (Chub and O'Donovan 2001). Is it possible that the transporters that set intracellular chloride concentration are more numerous or more active in the lidocaine-treated preparations, leading to a higher intracellular chloride concentration and consequently increasing the driving force for chloride?

We demonstrated that GABA mPSCs are depressed following an episode of SNA and progressively recover during the following IEI. In the lidocaine-treated preparations, GABA-mPSC amplitude was similarly depressed by the episode, but completely recovered in the shorter inter-episode intervals that are characteristic of the lidocaine-treated embryos. This led us to test the possibility that the modulation was influenced because the effectiveness of the chloride cotransporters had increased in the activity-reduced embryos. Because the modulation of GABAergic currents is likely determined by chloride cotransporters, we are now focusing on the possibility that an increase in chloride accumulation could underlie the increased rate of recovery of GABA mPSC amplitude in lidocaine preparations. We have therefore tested whether the reversal potential for GABA has become more depolarizing in the lidocaine embryos, increasing the driving force for these currents, and therefore GABAergic quantal amplitude.

Using perforated patch recordings, we have measured the GABA reversal potential by puffing on a $GABA_A$ agonist and blocking voltage-gated channels. In these experiments, we have indeed found that E_{GABA} is shifted from ~ -40 mV in control motoneurons to between -10 and -20 mV in motoneurons from lidocaine-treated embryos (Fig. 3.5). Thus, at least part of the increase in GABA quantal amplitude can be explained by the increase in driving force originating from the shift in E_{Gaba}. No such change has been described in activity-blocked, cultured neurons, although whole cell recordings were used and this may have obscured a contribution from changes in GABA reversal potential (Kilman et al. 2002). We have also begun to identify the cotransporters that are likely to mediate the chloride accumulation. We have made extracellular recordings of muscle nerves while puffing the $GABA_A$ agonist and measuring the potential generated. As expected, the response is reduced in both control and lidocaine-treated embryos when the NKCC1 cotransporter is blocked; however, it is only reduced to about 50%. This suggests that there are other means of chloride accumulation in addition to NKCC1. The findings suggest that one of the ways in which the homeostatic change in GABA quantal amplitude is achieved is through a change in the driving force for chloride. We have also identified that this change is likely mediated through changes in the function of chloride accumulators.

GABA in early embryos clearly has a profound influence on the development of these networks. We have shown that GABA transmission is modulated by SNA, and SNA is modulated by GABA transmission, and that GABAergic quantal amplitude increases in a homeostatic direction following reductions in SNA. It also appears that GABA signaling through its $GABA_A$ receptor is likely important in sensing

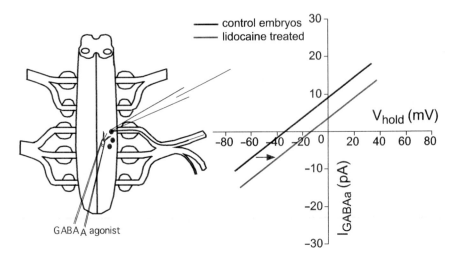

Fig. 3.5 E_{GABA} is shifted to a more depolarized level in activity-reduced embryos. Schematic shows the recording configuration for establishing the GABAergic reversal potential. Perforated patch recordings were made from motoneurons following $GABA_A$ agonist (isoguvacine) puffs, in the presence of voltage-gated channel blockers. Voltage ramps were made before and during $GABA_A$ agonist puffs to construct I–V plots in control and lidocaine-treated embryos. Approximate I–V plot is shown for control and treated embryos.

the activity levels that trigger homeostatic increases in quantal amplitude (Wilhelm and Wenner 2008). We have recently blocked $GABA_A$ transmission *in ovo* from E8-10, while leaving SNA largely intact. When this is done even, larger changes in quantal amplitude are observed, suggesting the importance of GABAergic transmission in the homeostatic process. The findings suggest that the compensatory changes are triggered when the network senses lowered GABA levels, as a proxy for activity.

These results suggest that spontaneous network activity can regulate the synaptic strength of motoneuron inputs during development, and that GABA plays a fundamentally important role in this process. Chronic reduction of activity or GABA signaling produces compensatory increases in both glutamatergic and depolarizing GABAergic mPSCs. Therefore, SNA appears to regulate the strength of network connections in a manner that could maintain levels of activity appropriate for proper limb development. By coordinately adjusting the strength of GABAergic and glutamatergic synaptic inputs as they face similar challenges (e.g., changes in cell size), this process could drive a balanced maturation of excitatory and inhibitory systems. Because of their shared depolarizing nature, certain features of excitatory and inhibitory synapses could be regulated in a mechanistically similar way, producing an initially coordinated development of the two systems. In this way, spontaneous network activity is likely to be important for the maturation of synaptic strength. SNA is particularly well suited to drive the maturation of synaptic inputs, because the great majority of spinal neurons, and their synapses, are recruited during episodes of SNA and this could allow the cells to measure the efficacy of their inputs. Future studies increasing SNA will be necessary to more completely determine the homeostatic nature of this synaptic regulation. Also, the spiking activity levels in spinal interneurons appear to be lower than that of motoneurons during episodes of SNA and the IEI (Chub and O'Donovan 2001; Ritter et al. 1999; Wenner and O'Donovan 2001). If synaptic strength is regulated in a similar way for interneurons, then we might expect compensatory responses to be triggered at different activity levels for different neurons. Because spontaneous network activity appears to occur in virtually every developing circuit, it is likely that this activity is important for synaptic maturation throughout the nervous system.

References

Akerman CJ, Cline HT (2007) Refining the roles of GABAergic signaling during neural circuit formation. Trends Neurosci 30:382–389

Abbott LF, and Nelson SB (2000) Synaptic plasticity: taming the beast. Nat Neurosci 3 Suppl, 1178–1183

Bekoff A, Stein P, Hamburger V (1975) Coordinated motor output in the hindlimb of the 7-day chick embryo. Proc Natl Acad Sci USA 72:1245–1248

Ben-Ari Y, Gaiarsa JL, Tyzio R, Khazipov R (2007) GABA: a pioneer transmitter that excites immature neurons and generates primitive oscillations. Physiol Rev 87:1215–1284

Bliss TV, and Lomo T (1973) Long-lasting potentiation of synaptic transmission in the dentate area of the anaesthetized rabbit following stimulation of the perforant path. J Physiol 232: 331–356

Burrone J, Murthy VN (2003) Synaptic gain control and homeostasis. Curr Opin Neurobiol 13:560–567

Chub N, O'Donovan MJ (1998) Blockade and recovery of spontaneous rhythmic activity after application of neurotransmitter antagonists to spinal networks of the chick embryo. J Neurosci 18:294–306

Chub N, O'Donovan MJ (2001) Post-episode depression of GABAergic transmission in spinal neurons of the chick embryo. J Neurophysiol 85:2166–2176

Chub N, Mentis GZ, O'Donovan MJ (2006) Chloride-sensitive MEQ fluorescence in chick embryo motoneurons following manipulations of chloride and during spontaneous network activity. J Neurophysiol 95:323–330

Cline HT (1991) Activity-dependent plasticity in the visual systems of frogs and fish. Trends Neurosci 14:104–111

Davis GW (2006) Homeostatic control of neural activity: from phenomenology to molecular design. Annu Rev Neurosci 29:307–323

Fedirchuk B, Wenner P, Whelan PJ, Ho S, Tabak J, O'Donovan MJ (1999) Spontaneous network activity transiently depresses synaptic transmission in the embryonic chick spinal cord. J Neurosci 19:2102–2112

Feller MB (1999) Spontaneous correlated activity in developing neural circuits. Neuron 22:653–656

Ge S, Goh EL, Sailor KA, Kitabatake Y, Ming GL, Song H (2006) GABA regulates synaptic integration of newly generated neurons in the adult brain. Nature 439:589–593

Gonzalez-Islas CE, Wenner P (2006) Spontaneous network activity in the embryonic spinal cord regulates AMPAergic and GABAergic synaptic strength. Neuron 49:563–575

Gu X, Spitzer NC (1997) Breaking the code: regulation of neuronal differentiation by spontaneous calcium transients. Dev Neurosci 19:33–41

Hall BK, Herring SW (1990) Paralysis and growth of the musculoskeletal system in the embryonic chick. J Morphol 206:45–56

Hamburger V (1977) The developmental history of the motor neuron. Neurosci Res Program Bull 15:1–37

Hanson MG, Landmesser LT (2004) Normal patterns of spontaneous activity are required for correct motor axon guidance and the expression of specific guidance molecules. Neuron 43:687–701

Hanson MG, Milner LD, Landmesser LT (2008) Spontaneous rhythmic activity in early chick spinal cord influences distinct motor axon pathfinding decisions. Brain Res Rev 57:77–85

Hawkins RD, Kandel ER, and Siegelbaum SA (1993) Learning to modulate transmitter release: themes and variations in synaptic plasticity. Annu Rev Neurosci 16:625–665

Jarvis C, Sutherland H, Mayne CN, Gilroy SJ, Salmons S (1996) Induction of a fast-oxidative phenotype by chronic muscle stimulation: mechanical and biochemical studies. Am J Physiol 270:C306–C312

Kandler K, Gillespie DC (2005) Developmental refinement of inhibitory sound-localization circuits. Trends Neurosci 28:290–296

Katz LC, Shatz CJ (1996) Synaptic activity and the construction of cortical circuits. Science 274:1133–1138

Kilman V, van Rossum MC, Turrigiano GG (2002) Activity deprivation reduces miniature IPSC amplitude by decreasing the number of postsynaptic $GABA_{(A)}$ receptors clustered at neocortical synapses. J Neurosci 22:1328–1337

LeMasson G, Marder E, Abbott LF (1993) Activity-dependent regulation of conductances in model neurons. Science 259:1915–1917

Linden DJ, and Connor JA (1995) Long-term synaptic depression. Annu Rev Neurosci 18:319–357

Lissin DV, Gomperts SN, Carroll RC, Christine CW, Kalman D, Kitamura M, Hardy S, Nicoll RA, Malenka RC, von Zastrow M (1998) Activity differentially regulates the surface expression of synaptic AMPA and NMDA glutamate receptors. Proc Natl Acad Sci USA 95:7097–7102

Malenka RC, and Nicoll RA (1993) NMDA-receptor-dependent synaptic plasticity: multiple forms and mechanisms. Trends Neurosci 16:521–527

Malenka RC, Nicoll RA (1999) Long-term potentiation–a decade of progress? Science 285:1870–1874

Marchetti C, Tabak J, Chub N, O'Donovan MJ, Rinzel J (2005) Modeling spontaneous activity in the developing spinal cord using activity-dependent variations of intracellular chloride. J Neurosci 25:3601–3612

Marder E, Goaillard JM (2006) Variability, compensation and homeostasis in neuron and network function. Nat Rev Neurosci 7:563–574

Miller KD (1994) Models of activity-dependent neural development.". Prog Brain Res 102:303–318

Milner LD, Landmesser LT (1999) Cholinergic and GABAergic inputs drive patterned spontaneous motoneuron activity before target contact. J Neurosci 19:3007–3022

O'Brien RJ, Kamboj S, Ehlers MD, Rosen KR, Fischbach GD, Huganir RL (1998) Activity-dependent modulation of synaptic AMPA receptor accumulation. Neuron 21:1067–1078

O'Donovan MJ (1987) In vitro methods for the analysis of motor function in the developing spinal cord of the chick embryo. Med Sci Sports Exerc 19:S130–S133

O'Donovan MJ (1999) The origin of spontaneous activity in developing networks of the vertebrate nervous system. Curr Opin Neurobiol 9:94–104

O'Donovan MJ, Chub N, Wenner P (1998) Mechanisms of spontaneous activity in developing spinal networks. J Neurobiol 37:131–145

Owens DF, Kriegstein AR (2002) Is there more to GABA than synaptic inhibition? Nat Rev Neurosci 3:715–727

Payne JA, Rivera C, Voipio J, Kaila K (2003) Cation-chloride co-transporters in neuronal communication, development and trauma. Trends Neurosci 26:199–206

Persson M (1983) The role of movements in the development of sutural and diarthrodial joints tested by long-term paralysis of chick embryos. J Anat 137:591–599

Rich MM, Wenner P (2007) Sensing and expressing homeostatic synaptic plasticity. Trends Neurosci 30:119–125

Ritter P, Wenner P, Ho S, Whelan PJ, O'Donovan MJ (1999) Activity patterns and synaptic organization of ventrally located interneurons in the embryonic chick spinal cord. J Neurosci 19:3457–3471

Roufa D, Martonosi AN (1981) Effect of curare on the development of chicken embryo skeletal muscle in ovo. Biochem Pharmacol 30:1501–1505

Shatz CJ (1990) Impulse activity and the patterning of connections during CNS development. Neuron 5:745–756

Singer JH, Berger AJ (2000) Development of inhibitory synaptic transmission to motoneurons. Brain Res Bull 53:553–560

Stent GS (1973) A physiological mechanism for Hebb's postulate of learning. Proc Natl Acad Sci U S A 70:997–1001

Tabak J, Senn W, O'Donovan MJ, Rinzel J (2000) Modeling of spontaneous activity in developing spinal cord using activity-dependent depression in an excitatory network. J Neurosci 20:3041–3056

Tabak J, Rinzel J, and O'Donovan MJ (2001) The role of activity-dependent network depression in the expression and self-regulation of spontaneous activity in the developing spinal cord. J Neurosci 21:8966–8978.

Turrigiano GG, Abbott LF, Marder E (1994) Activity-dependent changes in the intrinsic properties of cultured neurons. Science 264:974–977

Turrigiano GG, Leslie KR, Desai NS, Rutherford LC, Nelson SB (1998) Activity-dependent scaling of quantal amplitude in neocortical neurons. Nature 391:892–896

Turrigiano GG, Nelson SB (2004) Homeostatic plasticity in the developing nervous system. Nat Rev Neurosci 5:97–107

Turrigiano G (2007) Homeostatic signaling: the positive side of negative feedback. Curr Opin Neurobiol 17:318–324

Wenner P, O'Donovan MJ (2001) Mechanisms that initiate spontaneous network activity in the developing chick spinal cord. J Neurophysiol 86:1481–1498

Wilhelm J, Wenner P (2007) $GABA_A$ receptors: a potential sensor for homeostatic changes in GABAergic and AMPAergic synaptic strength in embryonic spinal cord. Soc for Neurosc Abstr 417:17

Yao H, Dan Y (2005) Synaptic learning rules, cortical circuits, and visual function. Neuroscientist 11:206–216

Yuste R, Nelson DA et al (1995) Neuronal domains in developing neocortex: mechanisms of coactivation. Neuron 14:7–17

Part II
Systems

Chapter 4
Regulation of Inhibitory Synapse Function in the Developing Auditory CNS

Dan H. Sanes, Emma C. Sarro, Anne E. Takesian, Chiye Aoki, and Vibhakar C. Kotak

The regulation of inhibitory synaptic strength begins during the period of synaptogenesis when the specificity of inhibitory and excitatory terminals becomes established. The mechanisms that underlie this process are just beginning to be understood. At the molecular level, the neuroligin family of cell adhesion molecules may selectively increase the formation of new inhibitory synapses (Chih et al. 2005; Levinson et al. 2005). However, experimental evidence from the auditory system suggests that establishing a balance between excitation and inhibition also depends on the selective elimination of inhibitory connections (Sanes and Siverls 1991; Sanes 1993; Gabriele et al. 2000a; Kapfer et al. 2002; Kim and Kandler 2003; Werthat et al. 2008; Franklin et al. 2008; Kandler et al. 2009) while similar pruning of excitatory terminals must also coexist.

The stabilization or elimination of a synapse can be influenced by neurotransmission itself, often referred to as *activity-dependent plasticity*. Although the activity-dependent modification of excitatory synapses has been the focus of intense study for many decades (see Sanes et al. 2006), there is a growing recognition that inhibitory synapses employ similar mechanisms. In fact, the strongest evidence for this view comes from experiments performed on the central auditory system (Sanes and Takács 1993; Gabriele et al. 2000b; Kapfer et al. 2002; Kim and Kandler 2003; Werthat et al. 2008; Franklin et al. 2008). Furthermore, the strength of inhibitory synapses is remarkably dynamic during development, even after the period when developing connections are eliminated. In this chapter, we explore how inhibitory synaptic gain is adjusted in the auditory CNS, especially during the period of postnatal maturation.

D.H. Sanes (✉), E.C. Sarro, A.E. Takesian, C. Aoki and V.C. Kotak
Center for Neural Science, New York University, 4 Washington Place, New York, NY, 10003, USA
e-mail: sanes@cns.nyu.edu

4.1 Spontaneous and Sound-Evoked Activity During Development

Many experiments demonstrating an influence of activity on inhibitory gain have been based on in vivo manipulations of the developing cochlea, often before the animal would hear airborne sound. Therefore, it is crucial to know the characteristics of neural activity in the developing auditory system at this time point, and whether manipulations of this sort can alter the normal amount or the pattern of synaptic transmission and action potentials. In this section, we review what is known about neural activity in the developing auditory CNS. However, there is an important caveat: there is not a single in vivo study that has measured spontaneous *inhibitory* synaptic activity during development, either in control animals or following a manipulation. This is important for studies that explored the effect of decreasing synaptic inhibition during development (Sanes et al. 1992; Sanes and Chokshi 1992; Moore 1992; Aponte et al. 1996; Kotak and Sanes 1996). Therefore, the precise functional impact of manipulations that "decrease" activity are yet to be determined.

Spontaneous action potentials have been recorded in central auditory regions before the onset of hearing, including in the gerbil cochlear nucleus and inferior colliculus (Woolf and Ryan 1985; Kotak and Sanes 1995). Much of this spontaneous activity arises in the periphery (Beutner and Moser 2001; Brandt et al. 2003; Tritsch et al. 2007; Jones et al. 2007), and it has been reported that cochlea removal or the blockade of action potentials with tetrodotoxin (TTX) leads to a complete cessation of spontaneous bursting activity in the embryonic chick cochlear nucleus (Lippe 1994). There is evidence for spontaneous activity in the auditory cortex that is independent of the cochlea, however. Oscillatory discharge has been observed in isolated thalamorecipient auditory cortex of gerbils during the first postnatal week (Kotak et al. 2007). Furthermore, calcium waves in isolated cortex are observed to sweep from caudal to rostral at rates of up to five waves per min, similar to those reported in other developing cortices (Garaschuk et al. 2000; Adelsberger et al. 2005). Importantly, such oscillations and calcium waves involve inhibitory transmission: focal delivery of GABA dampens bursting activity by hyperpolarizing the membrane potential as early as P3, while a $GABA_A$ receptor antagonist disrupts the synchronized cortical rhythms, leading to tonic discharge (Kotak et al. 2007).

As the auditory system matures, the effective sound level to elicit a response declines. Thus, high thresholds will initially limit the amount of sound-evoked activity, due largely to an immature auditory periphery (for review, see Fitzgerald and Sanes 2001). In gerbils, airborne sound can first elicit a response from the cochlea at about postnatal (P) day 12, and thresholds gradually decline to adult values by P30 (Woolf and Ryan 1984; McFadden et al. 1996). A similar developmental trajectory has been described for other rodent species (Romand 1992).

A second constraint on sound-driven neural activity is that dynamic range and maximum output are limited, both at the cochlea and within the CNS. In adult animals, central auditory neurons typically modulate their discharge rate over a 20–50 dB range of intensities, whereas animals perceive increments over approximately a 100 dB range. However, the dynamic range is quite limited at hearing onset. In gerbils, the cochlea encodes less than half of the adult sound level range at P12, and this input-output function has not fully matured at P30 (Woolf and Ryan 1984). Central auditory neurons appear to reflect this limitation in that maximum discharge rates display the same prolonged time course to reach adult values (Woolf and Ryan 1985; Sanes and Rubel 1988; Thornton et al. 1999).

The first studies to demonstrate a strong causal relationship between environmental stimulation and the development of connections were performed in the cat visual system. In these studies, decreasing visual stimulation during development led to a dramatic loss in the ability of the eyes to activate cortical neurons (Wiesel and Hubel 1965). More recently, spontaneous retinal activity has been shown to influence the refinement of retinal ganglion cell arbors within the superior colliculus during the first week after birth (Chandrasekaran et al. 2005). Experimental manipulations that injure or interfere with the cochlea generally reduce spontaneous and sound-driven electrical activity in the central auditory system (Bock and Webster 1974; Shepherd et al. 1999; Koerber et al. 1966; Tucci et al. 2001; Tucci et al. 1999; Lee et al. 2001; Cook et al. 2002), and thus can be used to study the effect of decreasing activity on the development of inhibitory synapse function.

4.2 Perturbation of Auditory System Activity Alters Inhibition

In the auditory cortex (ACx), inhibitory connections are outweighed by their excitatory counterparts by about 4:1, yet small deficiencies in inhibition can profoundly impact network properties (Chagnac-Amitai and Connors 1989; for review, see Fritschy and Brünig 2003). In the ACx activation of GABAergic circuits contributes to many response properties, including onset latencies, excitatory receptive fields, and motion processing (Müller and Scheich 1988; Horikawa et al. 1996; Chen and Jen 2000; Foeller et al. 2001; Firzlaff and Schuller 2001; Wang et al. 2002b). In the inferior colliculus, in vivo blockade of inhibitory synapses demonstrates that they contribute to a broad range of auditory coding properties (for reviews, see Pollak et al. 2002; Pollak et al. 2003).

Hearing impairments alter the coding properties within the central auditory system, and these changes may be explained, in part, by the alterations of inhibitory synaptic function described in this chapter. In particular, in vivo studies of deafened animals and of age-related hearing loss have suggested that adjustments of the strength of inhibitory afferents occur. For example, when animals are unilaterally deafened as neonates, acoustically-evoked activity in the ipsilateral inferior colliculus is increased in adults (Kitzes and Semple 1985; Szczepaniak and Moller 1995). Similar effects are observed following acute unilateral ablation in adult animals (McAlpine et al. 1997). Weakened sideband inhibition following noise and drug-induced hearing loss can also contribute to enhanced IC neuron discharge and expansion of frequency tuning curves (Wang et al. 2002a).

Because the effects of cochlear damage emerge rapidly, it has been suggested that excitatory inputs are "unmasked" by decreasing inhibitory drive from the deafened ear (Calford et al. 1993; Kimura and Eggermont 1999; Salvi et al. 2000; Norena et al. 2003).

For example, when a small section of the cochlea is damaged, the frequency tuning of cortical neurons expands due to the loss of surround inhibition (Rajan 1998). However, inhibitory inputs that are driven by the same frequencies as the excitatory inputs do not appear to decrease in strength (Rajan 2001). Thus, in vivo experiments suggest that inhibitory synaptic strength is altered after cochlear damage, but they cannot assess inhibitory synapses selectively. Many factors must be considered, including alterations to excitatory synapses and membrane properties within the many brain stem auditory nuclei. Nonetheless, direct assays of inhibitory markers or function can establish reduced or enhanced synaptic inhibition. As discussed below, there is now direct cellular and molecular evidence in support of this idea.

From a chronological perspective, the first indication that inhibitory synaptic properties were use-dependent came from studies on the CNS following hearing loss that is commonly observed during aging (Caspary et al. 2008). This work demonstrated a profound alteration of GABAergic properties in the inferior colliculus (Banay-Schwartz et al. 1989; Caspary et al. 1990). The research findings described in the following section establish that inhibitory gain is adjusted simultaneously at each level of the auditory CNS, although the first nuclei in the pathway tend to display non-homeostatic alterations. The cellular mechanisms by which inhibitory synaptic strength is set are quite diverse, and include both pre- and postsynaptic sites. In most cases, excitatory synaptic gain and membrane excitability are adjusted concurrently. In the following sections, we consider the evidence demonstrating how inhibitory synapses from ventral brainstem and midbrain to cortex are altered by experience.

4.3 Developmental Regulation of Inhibitory Synapses in the Lateral Superior Olive

We initially studied a group of inhibitory neurons that participate in a simple brain stem circuit that computes interaural level differences (ILD), a sound cue that is used to locate a sound along the horizontal axis. Lateral superior olivary (LSO) neurons each respond selectively to particular values of ILD by integrating excitatory inputs driven by the ipsilateral ear with inhibitory inputs driven by the contralateral ear (for review, see Tollin 2003). As shown in Fig. 4.1a, the inhibitory projection

---

Use-dependent long term depression of inhibitory synapses in the LSO was observed following low frequency stimulation (LFS) of the MNTB. MNTB-evoked maximum IPSPs were recorded in the absence and presence of the $GABA_B$ receptor antagonist, SCH-50911. In control neurons, synaptic depression was robust (43%) at 50–60 min following LFS (*filled circles*). Age-matched neurons treated with SCH-50911 (*open circles*) displayed an insignificant change in IPSP amplitude following LFS (mean ± SEM). (**c**) Following functional deafferentation of the MNTB by contralateral cochlear ablation (SNHL), evoked IPSPs were smaller (top traces and bar graph. In contrast, the unmanipulated ipsilateral excitatory pathway became stronger. Evoked EPSPs were longer in duration in SNHL neurons, and this was attributable to up-regulation of NMDA receptor function, as assessed with the NMDA receptor antagonist, AP-5 (bottom traces and bar graph). *N* values in bars

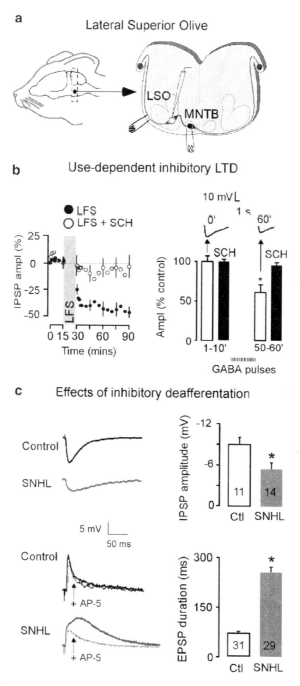

Fig. 4.1 Inhibitory synaptic plasticity in the developing LSO. (**a**) Schematic shows the position of the brain slice (*dashed box*) and location of LSO (*black circle*). The brain slice (*right panel*) contains the LSO, and the excitatory projection from the ipsilateral cochlear nucleus (*left stimulating electrode*) and the inhibitory projection from the MNTB (*right stimulating electrode*). (**b**)

originates from the medial nucleus of the trapezoid body (MNTB), and the excitatory projection from the cochlear nucleus (CN).

Projections from both the excitatory CN and the inhibitory MNTB form tonotopic maps in the LSO. As with other areas of the nervous system, these projections result from accurate outgrowth and innervation mechanisms, but there is also evidence that these projections undergo postnatal refinement through synapse elimination. In vivo recordings indicate that there is a significant improvement in the matching of excitatory and inhibitory sound frequencies between P13–14, when gerbils first respond to airborne sound, and adulthood (Sanes and Rubel 1988). Functional estimates of the number of excitatory and inhibitory terminals per LSO neuron suggest that convergence declines during development (Sanes 1993). Furthermore, single MNTB terminal arborizations in the LSO become physically restricted during development. During an early period of refinement, prior to the onset of hearing, the MNTB projection to the LSO undergoes a dramatic reduction in area (Kim and Kandler 2003; Kandler et al. 2009). During a subsequent period, after the onset of hearing, individual arbors are reduced by about 30%. The refinement of inhibitory MNTB arbors within LSO depends, in part, on their activity. When the contralateral cochlea is ablated at P7, single MNTB terminal arbors that were deafferented by the ablation fail to attain the normal level of anatomical specificity (Sanes and Takács 1993). A complementary phenomenon has also been described in a second target nucleus of the MNTB, the medial superior olivary nucleus (MSO). Terminals from the MNTB are eliminated from MSO dendrites during early development, and this process is prevented by unilateral cochlear ablation and diminished by rearing gerbils in white noise (Kapfer et al. 2002; Werthat et al. 2008). Finally, the projection from a GABAergic nucleus to the inferior colliculus also displays an anatomical refinement during development, and this can be reversed by cochlear albation (Gabriele et al. 2000a, 200b; Franklin et al. 2008).

There is a strong literature supporting a role for activity in the developmental elimination of excitatory synapses (for review, see Sanes et al. 2006). At the developing neuromuscular junction, activity-dependent excitatory synaptic depression is closely associated with the elimination of polyneuronal innervation (for review, see Wyatt and Balice-Gordon 2003). Therefore, an important question arising from these studies is whether the physical elimination of inhibitory synapses is associated with a weakening of inhibitory transmission. In fact, MNTB synapses do display a form of use-dependent long-term depression (LTD), and this form of inhibitory plasticity declines with age. Synapse depression is hypothesized to be an initial step in the elimination of excitatory synapses, possibly through the reduction of postsynaptic receptors (Li et al. 2001; Heynen et al. 2003). Therefore, inhibitory LTD could support synaptic remodeling in the developing LSO, and contribute to excitatory-inhibitory balance (Kotak and Sanes 2000).

Our studies of inhibitory LTD began with a rather unexpected observation. Inhibitory transmission within the LSO was thought of as exclusively glycinergic in adult animals (Moore and Caspary 1983; Sanes et al. 1987; Wenthold et al. 1987, 1990,), and we assumed that this held true for neonates. To our surprise, we found that inhibitory MNTB terminals are primarily GABAergic during the first postnatal week, before sound-evoked responses are present (Kotak et al. 1998; Korada and Schwartz 1999). This finding suggested that GABA release is necessary for the induction of inhibitory LTD, acting via G-protein coupled $GABA_B$ receptors. As shown in Fig. 4.1b (left), when whole-cell recordings were made from LSO neurons in the presence of a $GABA_B$ receptor antagonist (SCH-50911), we found that inhibitory LTD was almost completely eliminated (Kotak et al. 2001).

We next designed a specific test of whether postsynaptic $GABA_B$ receptor activation alone could induce depression. A micropipette, containing either GABA or glycine, was positioned in close proximity to the recorded LSO neuron. We found that focal delivery of GABA, but not glycine, was sufficient to trigger depression of the evoked hyperpolarizations (Chang et al. 2003). Furthermore, the GABA-induced depression could be blocked by the $GABA_B$ receptor antagonist (Fig. 4.1b). Together, these observations lend credibility to the notion that GABA plays a pivotal role in the induction and maintenance of inhibitory LTD (Kotak et al. 2001; Chang et al. 2003). The postsynaptic theory is consistent with our previous data that inhibitory LTD can be blocked by various kinds of intracellular manipulations exclusively in the recorded postsynaptic neuron (Kotak and Sanes 2000; Kotak and Sanes 2002). It should be noted that MNTB synapses in the LSO can also display long-term potentiation under specific stimulus conditions (Kotak and Sanes, unpublished observations). This property may provide an explanation for the enhanced conductance of the inhibitory synapses that become stabilized during the period of elimination in the developing rat LSO (Kim and Kandler 2003).

A second line of evidence for developmental homeostasis of inhibitory gain comes from studies in which the net inhibitory activity to LSO was experimentally decreased. In one set of experiments, one cochlea was surgically removed before the onset of hearing, which leads to the functional deafferentation of MNTB neurons. A second set of experiments used strychnine-containing continuous release pellets to attenuate the level of glycinergic transmission in vivo. As shown in Fig. 4.1c, manipulations of this sort influenced the maturation of synaptic properties. Whole-cell recordings showed that fewer LSO neurons received MNTB-evoked inhibition. In those neurons that did display MNTB-evoked IPSPs, the amplitude was significantly reduced, and this was accompanied by a depolarization in the IPSP reversal potential. More surprisingly, the unmanipulated ipsilateral pathway was altered dramatically: Ipsilaterally-evoked EPSPs were of much longer duration in experimental animals, and they were shortened significantly by an NMDA receptor antagonist, AP-5 (Kotak and Sanes 1996).

The effects of reducing inhibition during development may appear to support an anti-homeostatic mechanism: A down-regulation of inhibitory strength and an up-regulation of excitatory strength would not compensate for the manipulation. However, several observations suggest caution in drawing firm conclusions. For example, LSO neurons receive afferent projections from ipsilateral inhibitory and contralateral excitatory pathways (Wu and Kelly 1994; Kil et al. 1995). Furthermore, deafferentation elicits afferent sprouting, leading to novel innervation of MNTB and LSO (Kitzes et al. 1995; Russell and Moore 1995). Certain functional properties of the inhibitory MNTB projection must also be considered. For example, the MNTB projection releases both GABA and glycine during early development, and several reports demonstrate that the excitatory transmitter, glutamate, is also released (Kotak et al. 1998; Nabekura et al. 2004; Gillespie et al. 2005). Thus, the activation of metabotropic $GABA_B$ and glutamate receptors could play a primary role in regulating synaptic strength (Ene et al. 2003; Kotak and Sanes 1995; Kotak et al. 2001; Ene et al. 2007; Nishimaki et al. 2007).

4.4 Developmental Regulation of Inhibitory Synapse Gain in the Inferior Colliculus

Inhibitory projections to the inferior colliculus (IC) arise from many brain stem nuclei, and include both glycinergic and GABAergic afferents (for review, see Pollak et al. 2003). In a transverse brain slice preparation, much of the ascending inhibitory pathway can be activated with a stimulating electrode placed just ventral to the IC (Fig. 4.2a). We have used this approach to examine the effect of cochlear activity on the development of inhibitory synaptic transmission.

When gerbils are bilaterally deafened before the onset of hearing, evoked inhibitory postsynaptic potentials become much weaker than in control animals. We tested the ability of evoked IPSPs to block current-evoked action potentials, and found they were much less effective in deafened animals (Fig. 4.2b). In control neurons, the IPSPs block 97% of action potentials and the duration of inhibition lasted for 81 ms, but in deafened neurons only 43% of action potentials were blocked and the duration of inhibition was only 27 ms (Vale et al. 2003). Therefore, measures of synaptic strength indicate that inhibitory connections are less able to suppress suprathreshold events following deafness.

There are several changes that account for decreased inhibitory strength. The conductance of maximum evoked IPSCs is reduced by about 50% for all evoked IPSCs. This could be due to the loss of inhibitory afferents, a reduction in GABA or glycine release, a reduction of postsynaptic GABA or glycine receptors, or an alteration in the functional status of these receptors. There is also an alteration in release probability. When paired-pulses were delivered to the inhibitory pathway in control neurons, the evoked IPSCs exhibited facilitation. In contrast, paired-pulse facilitation is nearly eliminated in deafened animals (Vale and Sanes 2000).

The most dramatic change to inhibitory transmission involves the chloride (Cl⁻) battery. Synaptic inhibition elicited by $GABA_A$ or glycine receptor activation is mediated by a Cl⁻ conductance (Bormann et al. 1987). In most adult neurons, intracellular chloride $[Cl^-]_i$ is regulated by cation-chloride cotransporter family members: a Na–K–2Cl cotransporter (NKCC1) leads to cytoplasmic accumulation of chloride, and a K–Cl cotransporter (KCC2) extrudes chloride (Delpire et al. 1994; Payne et al. 1996;

neurons from control and SNHL animals. For control neurons, the mean E_{IPSC} was significantly more depolarized when the internal pipette solution contained Cs^+. However, for SNHL neurons, there was no significant difference between K^+- and Cs^+-containing pipettes. *n* values are in bars (*p 0.0001 vs K^+). Error bars indicate SEM

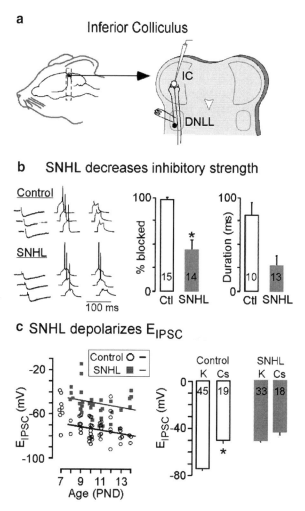

Fig. 4.2 Inhibitory synaptic plasticity in the developing IC. (**a**) Schematic shows the position of the brain slice (*dashed box*) and location of IC (*black circle*). The brain slice (*right*) contains the IC, and the ascending projection through the DNLL which includes both glycinergic and GABAergic afferents (*stimulating electrode*). (**b**) Traces show examples of evoked IPSPs, current-evoked action potentials, and the simultaneous presentation of IPSPs with action potentials (*from left to right*) in control and SNHL animals. The number of times that the evoked IPSP inhibited the AP in ten consecutive trials was counted and used to calculate the percentage of inhibition. Mean IPSP ability to inhibit APs is significantly lower in SNHL neurons compared with controls (*$p<0.0001$). Duration of inhibition was significantly shorter in BCA neurons compared with controls (**$p<0.005$). (**c**) IPSC reversal potential in gramicidin-perforated patch recordings with KCl in the internal pipette solution. SNHL caused a 24 mV depolarization in the mean E_{IPSC} (***$p<0.0001$ vs control), and the distribution of E_{IPSC} is plotted for neurons from control (*black circles*) and bilaterally deafened (*gray squares*) animals, along with regression lines (*left*). The effect of deafferentation on E_{IPSC} was apparent within 1 day of the surgical manipulation (at P7) and persisted during the age range studied (up to P14). The E_{IPSC} of control neurons at P7 is shown at the left. Bar graphs (*right*) show the effect of K^+ or Cs^+ in the recording pipette on E_{IPSC} in

Payne 1997; Payne et al. 2003). During early development [Cl⁻]$_i$ is relatively high due to NKCC1 activity (Plotkin et al. 1997; Clayton et al. 1998; Kanaka et al. 2001). As KCC2 expression increases, [Cl⁻]$_i$ drops below the electrochemical equilibrium (Lu et al. 1999; DeFazio et al., 2000; Hübner et al. 2001), leading to a transition from inhibitory synapse-evoked depolarizations to hyperpolarizations (Wang et al. 1994; Owens et al. 1996; Ehrlich et al. 1999; Kakazu et al. 1999; Rivera et al. 1999).

Using perforated patch recordings (which preserve the neurons' intracellular chloride concentration), we found that the mean IPSC reversal potential (E_{IPSC}) depolarized by 24 mV following hearing loss. As shown in Fig. 4.2c, this effect was present within one day of the in vivo manipulation and persisted at the longest interval examined, one week after deafening (Vale and Sanes 2000). The mechanisms responsible for E_{IPSC} depolarization are not yet fully understood, but many studies have shown that the expression of chloride transporter proteins account for the depolarizing inhibitory-evoked responses in immature neurons (Payne et al. 1996; Backus et al. 1998; Kazaku et al., 1999; Rivera et al. 1999; Williams et al. 1999; Balakrishnan et al. 2003; Vale et al. 2005; Blaesse et al. 2006). To examine the molecular basis of weakened inhibitory synapses in deaf animals, we measured the effect of three chloride transport blockers in control and deafened neurons. The results from one such experiment are shown in Fig. 4.2c. Control neurons displayed loss of chloride transport when challenged with intracellular cesium, whereas deaf neurons were relatively unaffected. Moreover, RT PCR and immunohistochemical analyses showed that KCC2 was expressed at normal levels in deaf neurons (Vale et al. 2003). Together, these results suggest that deafness disrupts the function of the chloride transporter without changing its expression.

It is estimated that over 90% of gerbil IC neurons are synaptically excited to spike threshold by contralateral sound stimulation, whereas only about 25% are so activated by the ipsilateral ear (Semple and Kitzes 1985; Brückner and Rübsamen 1995). These in vivo recordings suggest that the IC lobe contralateral to a deafened ear should be more deprived of excitatory input. If decreased postsynaptic activity leads to a down-regulation of inhibitory synaptic function, as suggested by the in vitro experiments cited above, then we would expect to observe a greater decrease of inhibitory strength contralateral to the deafened ear. This prediction is largely supported by our findings (Vale et al. 2004). Unilateral cochlear ablation led to a 23 mV depolarizing shift in the E_{IPSC} for IC neurons contralateral to the deafened ear, but only a 10 mV depolarization in the ipsilateral IC. Furthermore, commissural-evoked inhibitory synaptic conductance declined only contralateral to the ablated cochlea.

Our findings from developmentally deafened animals are consistent with findings from those deafened as adults. For example, there is a marked decrease in GABA release from within the IC following bilateral deafness in adult guinea pigs (Bledsoe et al. 1995). Furthermore, there is a profound loss of presynaptic GABA in the inferior colliculus, and a compensatory change in GABA$_A$ and GABA$_B$ receptor expression in very old animals, presumably due to age-related hearing loss (Caspary et al., 1995; Milbrandt et al. 1994, 1997). Although inhibition appears to be down-regulated in both young and old animals, the precise changes that occur may reflect the age of hearing loss, as well as the magnitude and duration of the loss (Suneja et al. 1998; Argence et al. 2006; Holt et al. 2005).

4.5 Developmental Regulation of Inhibitory Synapse Gain in the Auditory Cortex

Our understanding of auditory cortex (ACx) synaptic connectivity has improved somewhat during the past 5 years (for review, see Oswald et al. 2006). Thalamic stimulation in a brain slice preparation (schematized in Fig. 4.3a) typically evokes

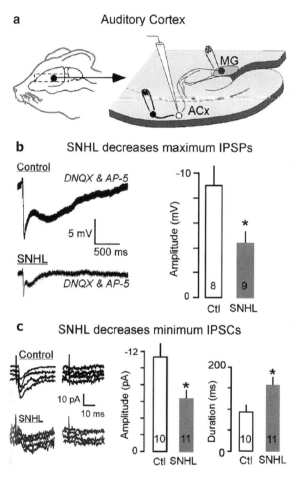

Fig. 4.3 Inhibitory synaptic plasticity in the developing ACx. (**a**) Schematic shows the position of the brain slice (*dashed box*) and location of ACx (*black circle*). The brain slice (*right*) contains the thalamorecipient ACx, the ascending projection from the thalamus (MG), and intracortical inhibitory projections (*bottom stimulating electrode*). (**b**) The maximum monosynaptic IPSP evoked by stimulating layer 2/3 is shown for control and SNHL neurons (*left*). These recordings were obtained in the presence of blockers of the ionotropic glutamate receptors DNQX and AP-5. The plot of putative monosynaptic IPSP amplitudes (*right*) shows a significant reduction for SNHL neurons. (**c**) Intracortical minimum evoked-IPSCs were recorded at −60 mV in the presence of ionotropic glutamate receptor blockers. The intensity at which minimum IPSCs (*left*) were discernible from failed responses (*right*) was then chosen for successive recordings. The amplitude of mean minimum evoked-IPSCs is smaller in SNHL neurons, while their mean duration is longer in SNHL neurons

a mixed excitatory-inhibitory response in layers 2–5, indicating the recruitment of feed forward GABAergic inhibition (Cruikshank et al. 2002). GABAergic neurons are distributed in all layers of ACx, and account for 15% of the ACx cells in gerbils (Foeller et al. 2001). Excitatory and inhibitory synaptic drive appear to be "balanced" insofar as their conductances are equivalent in magnitude (Wehr and Zador 2003; Tan et al. 2004).

Given the dramatic effects of disuse on the auditory brain stem, we were curious to explore the impact of deafness on inhibitory synapse function in the thalamorecipient ACx. Hearing loss was induced in gerbils just before the onset of hearing (P10), and synaptic function was subsequently assessed in a brain slice preparation (Fig. 4.3a). The maximum amplitude of intracortically-evoked GABAergic IPSPs was significantly smaller in deafened animals (Kotak et al. 2005) (Fig. 4.3b). As discussed for the IC (above), there could be many reasons for such a reduced response, including the death of inhibitory neurons. To determine whether individual inhibitory synapses produced smaller responses, we recorded spontaneous IPSCs and intracortically-evoked minimum amplitude IPSCs. The amplitudes of minimum-evoked IPSCs were significantly smaller while their durations were longer (Fig. 4.3c). A similar observation was made for spontaneous IPSCs, indicating the individual inhibitory terminals were weaker following hearing loss (Kotak et al. 2008). The longer duration of spontaneous and minimum-evoked IPSCs in ACx of deafened gerbils suggested that $GABA_A$ receptor subunit composition may have changed (Farrant and Kaila 2007). Therefore, we measured the pharmacosensitivity of two agonists, one specific for the α-1 subunit (zolpidem), and the other specific for the β-2/3 subunit (loreclezole) of the $GABA_A$ receptor. In control ACx neurons, each of these agonists enhanced the duration of spontaneous IPSCs, but this effect was absent following hearing loss (Fig. 4.4a). It is conceivable that the long IPSCs

---

prehearing neurons. Cumulative bar graphs show sIPSC duration before and after the application of agonist. Note that the agonist fails to prolong sIPSCs after hearing loss or before hearing onset. Number of neurons tested inside bars. The subunit agonist increased IPSC duration significantly only in control post-hearing animals ($X^2 = 6.8$; $p = 0.009$). (**b**) Electron micrographs show β2/3 subunit immunolabeling on the plasma membrane and at intracellular sites of layer 2/3 ACx pyramidal neurons, using the DAB procedure. In a control section (*left panel*), plasmalemmal labeling is apparent at a symmetric synapse (*arrowhead*) that extends intracellularly (*arrow*) within a distal segment of a dendrite from a control animal's auditory cortex. Such continuous labeling was tallied as labeled under both 'intracellular' and "plasmalemmal" categories. Nearby plasmalemmal labeling (*arrowhead* in the lower *right corner*) is less distinct and is associated with an axon terminal containing fewer vesicles. In a SNHL section (*right panel*), patches of intracellular labeling (*arrow*) within a dendrite of an SNHL animal's ACx. The patches are near a clearly unlabeled symmetric (presumably inhibitory) synapse. Arrowheads point to plasmalemmal immunolabeling. The bar graph (*bottom*) quantifies β2/3-DAB immunoreactivity from three controls and three SNHL animals. There was a significantly higher proportion of β2/3 immunolabeling on plasma membranes in controls (*white bars*), whereas intracellular β2/3 immunolabeling was greater in SNHL tissue (*black bars*). Asterisks indicate significance at $p<0.002$, determined by two tailed Student's *t*-test. *At* axon terminal, *D* Dendrite. Scale bars = 500 nm

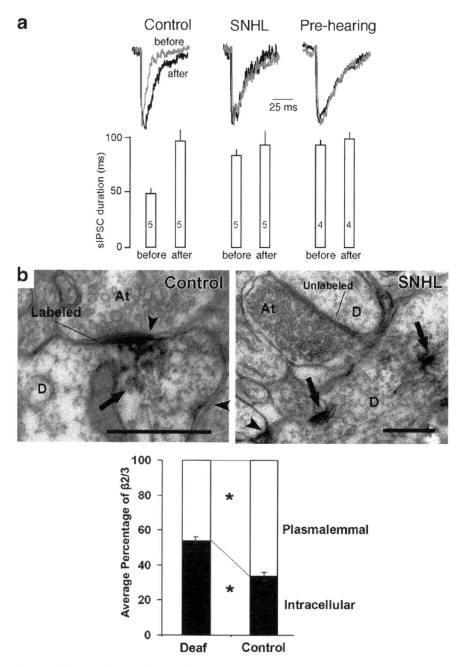

Fig. 4.4 Effects of hearing loss on $GABA_A$ receptors. (**a**) The top traces show representative IPSCs from those analyzed in control (*top left*), sensorineural hearing loss (SNHL, *top middle*), and pre-hearing neurons (*top right*). The gray trace is before the application of a [®]-2/3 subunit agonist, loreclezole, and the black trace is after application of the drug (traces normalized to control pre-drug amplitude). The agonist prolonged sIPSC duration in control, but not in SNHL or in

observed after hearing loss represent an immature phenotype and that hearing loss may delay the maturation of GABAergic transmission. To test this idea, the effect of each subunit agonist was determined for spontaneous IPSCs recorded at P10. Not only did the "pre-hearing" spontaneous IPSCs resemble those observed in much older deaf animals, but the $GABA_A$ receptor agonists did not prolong their duration, as seen in age-matched control neurons (Kotak et al. 2008).

We turned to quantitative EM-immunocytochemistry to explore whether these functional results could be explained, in part, by the localization of $GABA_AR$ subunits. As shown in Fig. 4.4b, the proportion β2/3 subunits declined significantly at the postsynaptic membrane in deaf neurons, and increased in the intracellular compartment just beneath the synapse (Sarro et al., 2006). This reduction was observed along pyramidal neuron somata, but not GABAergic interneurons. Thus, it appears that insertion and/or removal of $GABA_A$ receptors from the postsynaptic membrane was disrupted by hearing loss and this correlated with the reduced IPSC amplitude.

There are also clear signs that hearing loss alters release probability at inhibitory terminals. As shown in Fig. 4.5a, the frequency of spontaneous GABAergic IPSCs recorded in deaf neurons was over twice the rate recorded in controls. To better assess release probability, inhibitory short-term plasticity was examined using paired stimuli delivered intracortically in the presence of glutamate receptor antagonists. As shown in Fig. 4.5b, the IPSCs recorded in control neurons generally displayed paired-pulse facilitation (PPF), but this was significantly reduced in SNHL neurons (Takesian et al. 2007). This result is consistent with the elimination of inhibitory paired-pulse facilitation in the IC following hearing loss (Vale and Sanes 2000). We have recently found that a presynaptic GABAergic marker ($GAD_{65/67}$) increased by 47% in inhibitory terminals following hearing loss (Sarro et al. 2008); this may correlate with the sIPSC frequency increase observed in SNHL.

4.6 Summary

Our studies on the developing auditory CNS demonstrate that activity-dependent processes regulate inhibitory synaptic strength, a concept that has emerged from the work of several laboratories over the past 20 years. A broad set of analyses from normally developing animals and those with induced hearing loss reveal that inhibitory synapse function is adjusted at both pre- and postsynaptic loci. Following hearing loss, the net effect of these alterations leads to decreased inhibitory strength, as assessed by the ability of IPSPs to block action potentials (Fig. 4.2b).

Some key principles become apparent from these and related studies on inhibitory plasticity in the auditory CNS. First, it is clear that inhibitory gain is adjusted at each location of the ascending auditory system following hearing loss. Second, a decrease in inhibitory gain is often accompanied by a parallel increase in excitatory

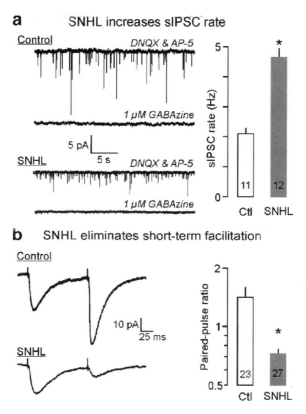

Fig. 4.5 Effect of hearing loss on GABA release. (**a**, *left*) Frequency of spontaneous (s) IPSCs is increased in SNHL neurons. sIPSCs recorded in the presence of ionotropic glutamate blockers (DNQX, AP-5) for 30s each in a P17 control neuron and an age-matched SNHL neuron are shown (*left panel*). Addition of 1 μM of GABAzine eliminates sIPSCs, showing that they are mediated by the activation of $GABA_A$ receptors. (**a**, *right*) Bar graph summarizing the mean frequencies of sIPSCs recorded from 11 control and 12 SNHL neurons. Note that the sIPSC frequency in SNHL neurons is significantly higher. (**b**) Paired pulse facilitation (PPF) of IPSCs is prevalent in control neurons, whereas SNHL neurons display paired pulse depression (PPD) (P17-21). Average of ten IPSCs evoked by paired extracellular stimuli recorded in the presence of ionotropic glutamate blockers (DNQX, AP-5) from control and SNHL neurons (*left*). Histogram of average PPR recorded from 21 control and 28 SNHL neurons (*log scale*) (*right*). Control neurons have a significantly greater PPR as compared with SNHL neurons (Chi square; $p<0.0001$). ISI = 120 ms, 20% above minimum stimulation

synaptic gain, and a rise in intrinsic firing. Together, these properties enhance neuronal excitability. Third, it is clear that activity-dependent regulation of inhibitory synaptic strength may depend on the age at which activity is disrupted, as well as the age when neuronal properties are examined. In the following ssection, we discuss how these adjustments emerge at each processing center in the ascending pathway, review some mechanisms that underlie inhibitory rescaling, and infer how decreased inhibitory gain could impact auditory processing.

4.6.1 Heirarchical Modification of Inhibitory Function

Following hearing loss, inhibitory gain has been shown to rescale at each relay station of the auditory CNS. In the cochlear nucleus (CN) of congenitally deaf mice, disrupted electrical activity reduces the amplitudes of miniature inhibitory currents (mIPSCs), increases single channel conductance carried by glycine receptors, and increases inhibitory postsynaptic sites assayed by gephyrin immunoreactivity (Leao et al. 2004a; 2004b). Similarly, in neomycin-deafened adult rats, there is a significant reduction of glycinergic presynaptic terminals in the cochlear nuclei and superior olivary complex (Asako et al. 2005; Buras et al. 2006). Furthermore, there is a concomitant increase in the amplitude of EPSCs due to increased probability of glutamate release (Oleskevich and Walmsley 2002). Thus, some level of activity-dependent homeostatic response is evident at the earliest stages of auditory processing (Burrone and Murthy 2003).

Although we do not yet know the rate of change following hearing loss in different regions of the auditory CNS, synaptic gain adjustments materialize rapidly, as early as 1 day following deafferentation. In the developing LSO, contralateral cochlear ablation (which leads to deafferentation of the inhibitory projection) produces a decline in inhibitory synaptic strength and an associated increase in excitatory strength (Fig. 4.1c). Thus, these gain adjustments occur in opposing directions to favor excitation.

Research on homeostatic control following perturbed activity in invertebrate and vertebrate systems led us to propose that hearing loss produces an imbalance in the currents that inhibit and excite auditory neurons (Marder et al., 2003; Marder and Goaillard 2006; Turrigiano 2007). In both the IC as well as the ACx, complete hearing loss triggered consistent transformations. First, maximum- and minimum-evoked IPSPs and IPSCs were reduced, supporting diminished inhibitory gain. Second, the kinetic properties of inhibitory currents in ACx failed to mature. Third, short-term inhibitory plasticity displayed depression, not facilitation, suggesting disrupted temporal processing. In concert, excitatory synaptic gain was scaled up. For example, larger amplitudes and durations of maximum and minimum-evoked, and spontaneous and miniature EPSCs imply heightened postsynaptic function (Vale and Sanes 2002; Kotak et al. 2005).

These findings agree with observations in other activity-deprived preparations. For example, a two-day visual deprivation period in early life elevates excitability within layer 4 by 25-fold, and this is associated with decreased inhibitory drive. Specifically, dual recordings show decreased strength of feedback inhibitory interneurons and increased strength of excitatory connections (Maffei et al. 2004).

Although we did not explore the effect of hearing loss on long-term inhibitory plasticity, one recent study showed compromised excitatory LTP and persistent LTD following hearing loss, suggesting plasticity may not develop properly in the absence of auditory experience (Kotak et al. 2007). Thus, activity-dependent inhibitory LTD and LTP observed in the normal LSO may exist at other auditory relays including the ACx.

In the following section, we consider several mechanisms that account for the regulation of inhibitory synaptic strength, especially following hearing loss.

4.6.2 Cellular Mechanisms that Regulate Inhibitory Gain

The activity-dependent mechanisms that operate at developing excitatory synapses have received much attention, particularly at motor neuron and retinal projections. Relatively less is known about the maturation of inhibitory synaptic function, particularly its regulation by activity-dependent mechanisms.

A broad range of cellular adjustments are initiated when inhibitory synapses are activated by a specific pattern, or when they are deafferented during normal development and aging (Morishita and Sastry 1991; Oda et al. 1995; 1998; Komatsu 1994; Caspary et al. 2005; Maffei et al. 2006). These studies highlight the fact that activity-dependent inhibitory scaling involves adjustment at both pre- and postsynaptic loci.

In normal developing gerbil LSO, our evidence shows that an activity-dependent reduction in inhibitory synaptic gain (LTD) is mediated by a rise in postsynaptic calcium and activation of specific kinases such as CaMKII, PKA and PKC. When each of these key intracellular factors was disrupted with specific postsynaptic antagonists, LTD was reduced or eliminated (Kotak and Sanes 2000; Kotak et al., 2002). This implies that inhibitory strength is adjusted by kinase-dependent phosphorylation of postsynaptic receptors. The LTD mechanism is engaged by GABA transmission (Fig. 4.1b). Specifically, postsynaptic $GABA_B$ receptor activation is necessary for LTD induction (Kotak et al. 2001; Chang et al. 2003).

A second postsynaptic mechanism that participates in inhibitory gain involves the regulation of intracellular chloride. Several lines of investigation showed that the chloride equilibrium potential becomes more negative during development (McCarthy et al. 2002). This is due to increased expression of KCC2 and reduced expression of NKCC1. Following hearing loss however, the IPSC reversal potential is depolarized. This is due to decreased KCC2 function, as assessed with perforated patch recordings, selective intracellular manipulations, and RT-PCR and immunocytochemistry of KCC2 (Vale et al. 2003).

Our third set of data supporting a postsynaptic mechanism was obtained from layer 2/3 pyramidal neurons of the developing ACx. Hearing loss results in a reduction of spontaneous and evoked IPSC amplitude, and this correlates with altered $GABA_A$ receptor subunit trafficking (Kotak et al. 2008; Sarro et al., 2006). Previous studies have shown that the number of $GABA_A$ receptors and their trafficking is activity-dependent (Tehrani and Barnes 1991; Barnes 1996; Paysan *et al.*, 1997; Nusser et al. 1997, 1998; Kilman et al. 2002). In our preparations, greater intracellular distribution and lesser postsynaptic membrane localization of the β2/3 subunit at symmetric synapses strongly supports the claim that hearing is vital for the proper mobilization and insertion of key $GABA_A$ receptor subunits. Further, the duration of IPSCs are determined, in part, by the specific $GABA_A$ receptor subunits expressed during development. For example, the two subunits we examined (α-1, β-2/3) play an obligatory role in agonist sensitivity and IPSC kinetics (Connolly et al. 1996; Baumann et al. 2002; Wisden et al. 1992; Amin and Weiss 1993; McKernan and Whiting 1996; for review, Möhler, 2006). When these subunits are upregulated during development, IPSC kinetics become faster; however, hearing loss prevented this transition, because IPSCs were not only long but the subunit-specific agonists failed to prolong them (Fig. 4.5a; Kotak et al. 2008).

Finally, presynaptic mechanisms may additionally regulate inhibitory strength. First, the increased mIPSC and sIPSC frequencies following hearing loss indicate augmented GABA release (Kotak et al. 2008). Second, an EM- immunocytochemical assay revealed an increase in the presynaptic GABAergic marker ($GAD_{65/67}$) (Sarro et al., 2006). Third, the diminution of paired-pulse facilitation following hearing loss, both in the midbrain and cortex, implies presynaptic change (Vale and Sanes 2000; Takesian et al. 2007; Fig. 4.5b). This result is consistent with results from similar studies on the visual cortex, in which visual deprivation leads to increased steady-state depression of IPSCs during trains of extracellular stimuli (Tang et al 2007) and during trains evoked by regular-spiking interneurons (Maffei et al 2004). Together, these findings suggest that GABA synthesis may be upregulated after *developmental* manipulations that decrease activity. However, age-related hearing loss is associated with a different set of alterations in the IC or ACx: there is a decrease in GABA-positive neurons, a decrease in GABA release, and an increase in GABA-mediated chloride influx (Caspary et a., 1990; Caspary et al. 1999; Ling et al. 2005). Thus, the inhibitory synaptic alterations that result from manipulations of activity are age-dependent.

If decreased cochlear activity leads to loss of inhibitory strength, then one would predict that increased activity would lead to up-regulation of inhibitory synapses. In fact, experimental induction of tinnitus (e.g., ringing in the ear) in rats leads to elevated GAD levels and increased $GABA_A$ receptor affinity in the IC (Bauer et al. 2000). It is possible that presynaptic GABA release is adjusted in response to the altered postsynaptic gain. Alternatively, postsynaptic gain may be a homeostatic response to presynaptic transmitter release. Further studies are needed to determine whether pre and postsynaptic gain adjustments are co-dependent.

In the following section, we consider the functional consequences of these cellular mechanisms and how they may influence auditory processing.

4.6.3 Effect of Inhibitory Gain on Auditory Processing

The cellular deficits we describe above may account for imbalanced acoustically-evoked discharge following hearing loss. In vivo recordings reveal robust modifications in auditory processing and reorganized tonotopy in the ACx following hair cell damage by ototoxic drugs, noise, or aging (Salvi et al. 2000; Syka, 2002; Caspary, 2005). Processing following hearing loss has been examined in cats using electrical stimulation with cochlear prosthetic devices, and the results are in broad agreement with our brain slice results. While some findings suggest that auditory deprivation leads to decreased synaptic drive, particularly to layer 5 (Klinke et al. 1999; Kral et al. 2000; 2009), there are also signs of increased excitability. Specifically, electrode-evoked thresholds are lower, spatial tuning curves are broader, and cochleotopy appears imprecise (Raggio and Schreiner 1999, 2003). These researchers suggest that diminished cortical inhibition could explain some of these characteristics. Their in vivo recordings do not distinguish between changes in the brainstem (Snyder et al. 2000; Moore et al. 2002) and those that have occurred locally in the cortex. Extracellular potentials also cannot tell us which cortical synapses or intrinsic properties have been altered, and in what manner.

Even unilateral deafferentation induces an increase in sound-evoked activity in the ipsilateral IC and ACx, and such properties may reflect an imbalanced interaction between inhibition and excitation (Kitzes and Semple 1985; McAlpine et al. 1997; Mossop et al. 2000). Following unilateral deafness, the ipsilateral IC exhibits far more excitatory responses than normal, suggesting that inhibition has been weakened (Kitzes and Semple 1985 ; McAlpine et al. 1997). In fact, there is a marked decrease in GABA release from the Central nucleus of the IC (CIC) following bilateral deafness (Bledsoe et al. 1995; but see Suneja et al. 1998). Our experiment on unilaterally deafened animals suggests that decreased inhibition following hearing loss may explain such an imbalance (Vale et al. 2004).

Inhibitory short-term plasticity shifts significantly from a facilitating to a depressing mode following hearing loss (Fig. 4.5b), and this is in agreement with a previously reported shift of inhibitory short-term plasticity in the IC (Vale and Sanes 2000). In the visual cortex, deprivation leads to increased steady-state depression of IPSCs during trains of extracellular stimuli (Tang et al 2007) and during trains evoked by regular-spiking interneurons (Maffei et al 2004).

During age-related hearing loss (i.e., presbycusis), there is a profound loss of presynaptic GABA in the inferior colliculus, and an associated change in $GABA_A$ and $GABA_B$ receptor expression (Caspary et al., 1995; Milbrandt et al. 1994, 1997). Each of these changes may contribute to age-related deficits in performance on auditory tasks. For example, in the dorsal cochlear nucleus, aging fusiform neurons respond with a greater maximum discharge to tones than those recorded from young adults; this finding is consistent with an age-related loss of glycinergic inhibition. Therefore,- clinically observed age-related central sensory processing deficits may be attributable to compromised function of inhibitory synapses (Caspary et al. 2005).

Following cochlear trauma, there are also signs from in vivo recordings that decreased inhibition may contribute to processing deficits. For example, spontaneous action potentials in ACx increase following cochlear trauma (Salvi et al. 2000; Wang et al. 2002a; Norena and Eggermont 2003; Seki and Eggermont 2003). The decreased inhibition discussed above could account, in part, for these in vivo changes (Kotak et al. 2005, 2008).

Using in vivo manipulations and whole-cell recordings in auditory brain slices in combination, we have directly assessed the mechanisms that govern inhibitory synaptic strength. Together, these findings suggest that the perceptual deficits that attend hearing loss are not solely attributable to peripheral factors, as is often assumed. Hearing loss-induced alterations to CNS inhibitory synapses must now be considered as a principal basis for diminished behavioral performance. Furthermore, these cellular findings could offer clues to the design of strategies for ameliorating the effects of early hearing loss. It may be possible to restore compromised auditory deficits in the hearing impaired by drugs that potentiate $GABA_A$ receptor function. For example, gap detection thresholds are elevated in aging gerbils, but normal performance can be rescued with a drug that elevates GABA levels (Gleich et al. 2003). Similarly, orientation and direction sensitivity of neurons in the visual cortex of aging primates can be reinstated by the administration of GABA agonists (Leventhal et al., 2000). We propose that processing deficits following hearing loss are, in part, due to diminished inhibitory strength along the entire ascending auditory pathway. Therefore, synaptic inhibition is a plausible candidate mechanism for clinical interventions to enhance perceptual skills.

Acknowledgments NIH DC006864 (DHS and VCK), DC008920 (AET), DC009729 (ECS), and NS41091, EY13145, P30 EY13079, DA009618-09 (CA)

References

Adelsberger H, Garaschuk O, Konnerth A (2005) Cortical calcium waves in resting newborn mice. Nat Neurosci 8:988–990
Aponte JE, Kotak VC, Sanes DH (1996) Decreased synaptic inhibition leads to dendritic hypertrophy prior to the onset of hearing. Auditory Neurosci 2:235–240
Argence M, Saez I, Sassu R, Vassias I, Vidal PP, de Waele C (2006) Modulation of inhibitory and excitatory synaptic transmission in rat inferior colliculus after unilateral cochleectomy: an in situ and immunofluorescence study. Neuroscience 141:1193–1207
Amin J, Weiss DS (1993) $GABA_A$ receptor needs two homologous domains of the beta-subunit for activation by GABA but not by pentobarbital. Nature 366:565–569
Asako M, Holt AG, Griffith RD, Buras ED, Altschuler RA (2005) Deafness-related decreases in glycine-immunoreactive labeling in the rat cochlear nucleus. J Neurosci Res 81:102–109
Backus KH, Deitmer JW, Friauf E (1998) Glycine-activated currents are changed by coincident membrane depolarization in developing rat auditory brainstem neurones. J Physiol (Lond) 507:783–794
Balakrishnan V, Becker M, Löhrke S, Nothwang HG, Güresir E, Friauf E (2003) Expression and function of chloride transporters during development of inhibitory neurotransmission in the auditory brainstem. J Neurosci 23:4134–4145
Banay-Schwartz M, Lajtha A, Palkovits M (1989) Changes with aging in the levels of amino acids in rat CNS structural elements I. Glutamate and related amino acids. Neurochem Res 14:555–562
Barnes EM Jr (1996) Use-dependent regulation of $GABA_A$ receptors. Int Rev Neurobiol 39:53–76
Baumann SW, Baur R, Sigel E (2002) Forced subunit assembly in $\alpha1\beta2\gamma2$ $GABA_A$ receptors. J Biol Chem 277:46020–46025
Blaesse P, Guillemin I, Schindler J, Schweizer M, Delpire E, Khiroug L, Friauf E, Nothwang HG (2006) Oligomerization of KCC2 correlates with development of inhibitory neurotransmission. J Neurosci 26:10407–10419
Bauer CA, Brozoski TJ, Holder TM, Caspary DM (2000) Effects of chronic salicylate on GABAergic activity in rat inferior colliculus. Hear Res 147:175–182

Beutner D, Moser T (2001) The presynaptic function of mouse cochlear inner hair cells during development of hearing. J Neurosci 21:4593–4599

Bledsoe SC Jr, Nagase S, Miller JM, Altschuler RA (1995) Deafness-induced plasticity in the mature central auditory system. NeuroReport 7:225–229

Bock GR, Webster WR (1974) Spontaneous activity of single units in the inferior colliculus of anesthetized cats. Brain Res 76:150–154

Bormann J, Hamill OP, Sakmann B (1987) Mechanism of anion permeation through channels gated by glycine and γ-aminobutyric acid in mouse cultured spinal neurones. J Physiol (Lond) 385:243–286

Brandt A, Striessnig J, Moser T (2003) CaV1.3 channels are essential for development and presynaptic activity of cochlear inner hair cells. J Neurosci 23:10832–10840

Brückner S, Rübsamen R (1995) Binaural response characteristics in isofrequency sheets of the gerbil inferior colliculus. Hear Res 86:1–14

Buras ED, Holt AG, Griffith RD, Asako M, Altschuler RA (2006) Changes in glycine immunoreactivity in the rat superior olivary complex following deafness. J Comp Neurol 494:179–189

Burrone J, Murthy VN (2003) Synaptic gain control and homeostatsis. Curr Opinion Neurobiol 13:560–567

Calford MB, Rajan R, Irvine DR (1993) Rapid changes in the frequency tuning of neurons in cat auditory cortex resulting from pure-tone-induced temporary threshold shift. Neuroscience 55:953–964

Caspary DM, Raza A, Lawhorn Armour BA, Pippin J, Arneric SP (1990) Immunocytochemical and neurochemical evidence for age-related loss of GABA in the inferior colliculus: implications for neural presbycusis. J Neurosci 10:2363–2372

Caspary DM, Holder TM, Hughes LF, Milbrandt JC, McKernan RM, Naritoku DK (1999) Age-related changes in GABA(A) receptor subunit composition and function in rat auditory system. Neuroscience 93:307–312

Caspary DM, Schatteman TA, Hughes LF (2005) Age-related changes in the inhibitory response properties of dorsal cochlear nucleus output neurons: role of inhibitory inputs. J Neurosci 25:10952–10959

Caspary DM, Milbrandt JC, Helfert RH (1995) Central auditory aging: GABA changes in the inferior colliculus. Exp Gerontol 30:349-360.

Caspary DM, Ling L, Turner JG, Hughes LF (2008) Inhibitory neurotransmission, plasticity and aging in the mammalian central auditory system. J Exp Biol 211:1781–1791

Chagnac-Amitai Y, Connors BW (1989) Horizontal spread of synchronized activity in neocortex and its control by GABA-mediated inhibition. J Neurophysiol 61:747–758

Chandrasekaran AR, Plas DT, Gonzalez E, Crair MC (2005) Evidence for an instructive role of retinal activity in retinotopic map refinement in the superior colliculus of the mouse. J Neurosci 25:6929–6938

Chang EH, Kotak VC, Sanes DH (2003) Long-term depression of synaptic inhibition is expressed postsynaptically in the developing auditory system. J Neurophysiol 90:1479–1488

Chen QC, Jen PH (2000) Bicuculline application affects discharge patterns, rate-intensity functions, and frequency tuning characteristics of bat auditory cortical neurons. Hear Res 150:161–174

Chih B, Engelman H, Scheiffele P (2005) Control of excitatory and inhibitory synapse formation by neuroligins. Science 307:1324–1328

Clayton GH, Owens GC, Wolff JS, Smith RL (1998) Ontogeny of cation-Cl⁻ cotransporter expression in rat neocortex. Brain Res Dev Brain Res 109:281–292

Connolly CN, Wooltorton JR, Smart TG, Moss SJ (1996) Subcellular localization of gamma-aminobutyric acid type A receptors is determined by receptor beta subunits. Proc Natl Acad Sci USA 93:9899–9904

Cook RD, Hung TY, Miller RL, Smith DW, Tucci DL (2002) Effects of conductive hearing loss on auditory nerve activity in gerbil. Hear Res 164:127–137

Cruikshank SI, Rose HJ, Metherate R (2002) Auditory thalamocrotical synaptic transmission in vitro. J Neurophysiol 87:361–384

Delpire E, Rauchman MI, Beier DR, Hebert SC, Gullans SR (1994) Molecular cloning and chromosome localization of a putative basolateral Na(+)-K(+)-2Cl(-) cotransporter from mouse inner medullary collecting duct (mIMCD-3) cells. J Biol Chem 269:25677–25683

DeFazio RA, Keros S, Quick MW, Hablitz JJ (2000) Potassium-couple chloride cotransport controls intracellular chloride in rat neocortical pyramidal neurons. J Neurosci 20:8069-8076.

Ehrlich I, Lohrke S, Friauf E (1999) Shift from depolarizing to hyperpolarizing glycine action in rat auditory neurones is due to age-dependent Cl⁻ regulation. J Physiol 520:121–137

Ene FA, Kalmbach A, Kandler K (2007) Metabotropic glutamate receptors in the lateral superior olive activate TRP-like channels: age- and experience-dependent regulation. J Neurophysiol 97:3365–3375

Ene FA, Kullmann PH, Gillespie DC, Kandler K (2003) Glutamatergic calcium responses in the developing lateral superior olive: receptor types and their specific activation by synaptic activity patterns. J Neurophysiol 90:2581–2591

Farrant M, Kaila K (2007) The cellular, molecular and ionic basis of GABA(A) receptor signalling. Prog Brain Res 160:59–87

Firzlaff U, Schuller G (2001) Motion processing in the auditory cortex of the rufous horseshoe bat: role of GABAergic inhibition. Eur J Neurosci 14.1687–1701

Fitzgerald KK, Sanes DH (2001) The development of stimulus coding in the auditory system. In: Jahn E, Santos-Sacchi J (eds) Physiology of the Ear, 2nd edn. San Diego, Singular Publishing

Foeller E, Vater M, Kössl M (2001) Laminar analysis of inhibition in the gerbil primary auditory cortex. J Assoc Res Otolaryngol 2:279–296

Fritschy JM, Brünig I (2003) Formation and plasticity of GABAergic synapses: physiological mechanisms and pathophysiological implications. Pharmacol Ther 98:299–323

Franklin SR, Brunso-Bechtold JK, Henkel CK (2008) Bilateral cochlear ablation in postnatal rat disrupts development of banded pattern of projections from the dorsal nucleus of the lateral lemniscus to the inferior colliculus. Neurosci 154:346–354

Gabriele ML, Brunso-Bechtold JK, Henkel CK (2000a) Development of afferent patterns in the inferior colliculus of the rat: projection from the dorsal nucleus of the lateral lemniscus. J Comp Neur 416:368–382

Gabriele ML, Brunso-Bechtold JK, Henkel CK (2000b) Plasticity in the development of afferent patterns in the inferior colliculus of the rat after unilateral cochlear ablation. J Neurosci 20:6939–6949

Garaschuk O, Linn J, Eilers J, Konnerth A (2000) Large-scale oscillatory calcium waves in the immature cortex. Nat Neurosci 3:452–9

Gillespie DC, Kim G, Kandler K (2005) Inhibitory synapses in the developing auditory system are glutamatergic. Nat Neurosci 8:332–338

Gleich O, Hamann I, Klump GM, Kittel M, Strutz J (2003) Boosting GABA improves impaired auditory temporal resolution in the gerbil. NeuroReport 14:1877–1880

Heynen AJ, Yoon BJ, Liu CH, Chung HJ, Huganir RL, Bear MF (2003) Molecular mechanism for loss of visual cortical responsiveness following brief monocular deprivation. Nat Neurosci 6:854–862

Holt AG, Asako M, Lomax CA, MacDonald JW, Tong L, Lomax MI, Altschuler RA (2005) Deafness-related plasticity in the inferior colliculus: gene expression profiling following removal of peripheral activity. J Neurochem 93:1069–1086

Horikawa J, Hosokawa Y, Kubota M, Nasu M, Taniguchi I (1996) Optical imaging of spatiotemporal patterns of glutamatergic excitation and GABAergic inhibition in the guinea-pig auditory cortex in vivo. J Physiol 497:629–638

Hübner CA, Stein V, Hermans-Borgmeyer I, Meyer T, Ballanyi K, Jentsch TJ (2001) Disruption of KCC2 reveals an essential role of K–Cl cotransport already in early synaptic inhibition. Neuron 30:515–524

Jones TA, Leake PA, Snyder RL, Stakhovskaya O, Bonham B (2007) Spontaneous discharge patterns in cochlear spiral ganglion cells before the onset of hearing in cats. J Neurophysiol 98:1898–1908

Kakazu Y, Akaike N, Komiyama S, Nabekura J (1999) Regulation of intracellular chloride by cotransporters in developing lateral superior olive neurons. J Neurosci 19:2843–2851

Kanaka C, Ohno K, Okabe A, Kuriyama K, Itoh T, Fukuda A, Sato K (2001) The differential expression patterns of messenger RNAs encoding K-Cl cotransporters (KCC1, 2) and Na-K-2Cl cotransporter (NKCC1) in the rat nervous system. Neuroscience 104:933–946

Kandler K, Clause A, Noh J (2009) Tonotopic reorganization of developing auditory brainstem circuits. Nat Neurosci 12:711–717

Kapfer C, Seidl AH, Schweizer H, Grothe B (2002) Experience-dependent refinement of inhibitory inputs to auditory coincidence-detector neurons. Nat Neurosci 5:247–253

Kazaku I, Akaike N, Komiyama S, Nabekura J (1999) Regulation of intracellular chloride by cotransporters in developing lateral superior olive neurons. J Neurosci 19:2843–2851.

Kil J, Kageyama GH, Semple MN, Kitzes LM (1995) Development of ventral cochlear nucleus projections to the superior olivary complex in gerbil. J Comp Neurol 353:317–340

Kilman V, van Rossum MC, Turrigiano GG (2002) Activity deprivation reduces miniature IPSC amplitude by decreasing the number of postsynaptic GABAA receptors clustered at neocortical synapses. J Neurosci 22:1328–1337

Kim G, Kandler K (2003) Elimination and strengthening of glycinergic/GABAergic connections during tonotopic map formation. Nat Neurosci 6:282–290

Kimura M, Eggermont JJ (1999) Effects of acute pure tone induced hearing loss on response properties in three auditory cortical fields in cat. Hear Res 135:146–162

Kitzes LM, Kageyama GH, Semple MN, Kil J (1995) Development of ectopic projections from the ventral cochlear nucleus to the superior olivary complex induced by neonatal ablation of the contralateral cochlea. J Comp Neurol 353:341–363

Kitzes LM, Semple MN (1985) Single-unit responses in the inferior colliculus: effects of neonatal unilateral cochlear ablation. J Neurophysiol 53:1483–1500

Klinke R, Kral A, Heid S, Tillein J, Hartmann R (1999) Recruitment of the auditory cortex in congenitally deaf cats by long-term cochlear electrostimulation. Science 285:1729–1733

Koerber KC, Pfeiffer RR, Warr WB, Kiang NY (1966) Spontaneous spike discharges from single units in the cochlear nucleus after destruction of the cochlea. Exp Neurol 16:119–130

Komatsu Y (1994) Age-dependent long-term potentiation of inhibitory synaptic transmission in rat visual cortex. J Neurosci 14:6488–6499

Korada S, Schwartz IL (1999) Development of GABA, glycine and their receptors in the auditory brainstem of gerbil: a light and electron microscopic study. J Comp Neurol 409:664–681

Kotak VC, Sanes DH (1995) Synaptically evoked prolonged depolarizations in the developing auditory system. J Neurophysiol 74:1611–1620

Kotak VC, Sanes DH (1996) Developmental influence of glycinergic inhibition: Regulation of NMDA- mediated EPSPs. J Neurosci 16:1836–1843

Kotak VC, Korada S, Schwartz IR, Sanes DH (1998) A developmental shift from GABAergic to glycinergic transmission in the central auditory system. J Neurosci 18:4646–4655

Kotak VK, Sanes DH (2000) Long-Lasting Inhibitory Synaptic Depression is Age- and Calcium-Dependent. J Neurosci 20:5820–5826

Kotak VC, DiMattina C, Sanes DH (2001) $GABA_B$ and Trk receptor signaling mediates long-lasting inhibitory synaptic depression. J Neurophysiol 86:536–540

Kotak VC, Sanes DH (2002) Postsynaptic kinase signaling underlies inhibitory synaptic plasticity. J Neurobiol 53:36–43

Kotak VC, Fujisawa S, Lee FA, Karthikeyan O, Aoki C, Sanes DH (2005) Hearing loss raises excitability in the auditory cortex. J Neurosci 25:3908–3918

Kotak VC, Sadahiro M, Fall CP (2007) Developmental expression of endogenous oscillations and waves in the auditory cortex involves calcium, gap junctions, and GABA. Neurosci 146:1629–1639

Kotak VC, Takesian AE, Sanes DH (2008) Hearing loss prevents the maturation of GABAergic transmission in the auditory cortex. Cereb Cortex 18:2098–2108

Kral A, Hartmann R, Tillein J, Heid S, Klinke R (2000) Congenital auditory deprivation reduces synaptic activity within the auditoy cortex in a layer specific manner. Cereb Cortex 10:714–726

Kral A, Tillein J, Peter Hubka P, Schiemann D, Heid S, Hartmann R (2009) Spatiotemporal patterns of cortical activity with bilateral cochlear implants in congenital deafness. J Neurosci 29:811–827

Leao RN, Berntson A, Forsythe ID, Walmsley B (2004a) Reduced low-voltage activated K^+ conductances and enhanced central excitability in a congenitally deaf (dn/dn) mouse. J Physiol 559:25–33

Leao RN, Oleskevich S, Sun H, Bautista M, Fyffe RE, Walmsley B (2004b) Differences in glycinergic mIPSCs in the auditory brain stem of normal and congenitally deaf neonatal mice. J Neurophysiol 91:1006–1012

Lee DS, Lee JS, Oh SH, Kim SK, Kim JW, Chung JK, Lee MC, Kim CS (2001) Cross-modal plasticity and cochlear implants. Nature 409:149–150

Leventhal AG, Youngchang W, Mingliang Pu, Yifeng Z, Yuanye M (2003) GABA and its agonists improved visual cortical function in senescent monkeys. Science 300:812-815.

Levinson JN, Chery N, Huang K, Wong TP, Gerrow K, Kang R, Prange O, Wang YT, El-Husseini A (2005) Neuroligins mediate excitatory and inhibitory synapse formation: involvement of PSD-95 and neurexin-1beta in neuroligin-induced synaptic specificity. J Biol Chem 280:17312–17319

Li MX, Jia M, Jiang H, Dunlap V, Nelson PG (2001) Opposing actions of protein kinase A and C mediate Hebbian synaptic plasticity. Nat Neurosci 4.871–872

Ling LL, Hughes LF, Caspary DM (2005) Age-related loss of the GABA synthetic enzyme glutamic acid decarboxylase in rat primary auditory cortex. Neuroscience 132:1103–1113

Lippe WR (1994) Rhythmic spontaneous activity in the developing avian auditory system. J Neurosci 14:1486–1495

Lu J, Karadhed M, Delpire E (1999) Developmental regulation of the neuronal-specific isoform of K–Cl cotranspoter KCC2 in postnatal rat brains. J Neurobiol 39:558–568

Maffei A, Nelson SB, Turrigiano GG (2004) Selective reconfiguration of layer 4 visual cortical circuitry by visual deprivation. Nat Neurosci 7:1353–1359

Maffei A, Nataraj K, Nelson SB, Turrigiano GG (2006) Potentiation of cortical inhibition by visual deprivation. Nature 443:81–84

Marder E, Goaillard JM (2006) Variability, compensation and homeostasis in neuron and network function. Nat Rev Neurosci 7:563–574

McCarthy MM, Auger AP, Perrot-Sinal TS (2002) Getting excited about GABA differences in the brain. Trends Neurosci 25:307–312

McAlpine D, Martin RL, Mossop JE, Moore DR (1997) Response properties of neurones in the inferior colliculus of the monaurally-deafened ferret to acoustic stimulation of the intact ear. J Neurophysiol 78:767–779

McFadden SL, Walsh EJ, McGee J (1996) Onset and development of auditory brainstem responses in the Mongolian gerbil (*Meriones unguiculatus*). Hear Res 100:68–79

McKernan RM, Whiting PJ (1996) Which $GABA_A$-receptor subtypes really occur in the brain? Trends Neuroscience 19:139–143

Milbrandt JC, Albin RL, Caspary DM (1994) Age-related decrease in GABAB receptor binding in the Fischer 344 rat inferior colliculus. Neurobiol Aging 15:699–703

Milbrandt JC, Hunter C, Caspary DM (1997) Alterations of GABAA receptor subunit mRNA levels in the aging Fischer 344 rat inferior colliculus. J Comp Neurol 379:455–465

Möhler H (2006) $GABA_A$ receptor diversity and pharmacology. Cell Tissue Res 326:505–516

Moore DR (1992) Trophic influences of excitatory and inhibitory synapses on neurones in the auditory brain stem. Neuro Report 3:269–272

Moore MJ, Caspary DM (1983) Strychnine blocks binaural inhibition in lateral superior olivary neurons. J Neurosci 3:237–42

Moore CM, Vollmer M, Leake PA, Snyder RL, Rebscher SJ (2002) The effects of chronic intracochlear electrical stimulation on inferior colliculus spatial representation in adult deafened cats. Hear Res 164:82–96

Morishita W, Sastry BR (1991) Chelation of postsynaptic Ca^{2+} facilitates long-term potentiation of hippocampal IPSPs. NeuroReport 2:533–6

Mossop JE, Wilson MJ, Caspary DM, Moore DR (2000) Down-regulation of inhibition following unilateral deafening. Hear Res 147:183–187

Müller CM, Scheich H (1988) Contribution of GABAergic inhibition to the response characteristics of auditory units in the avian forebrain. J Neurophysiol 59:1673–1689

Nabekura J, Katsurabayashi S, Kakazu Y, Shibata S, Matsubara A, Jinno S, Mizoguchi Y, Sasaki A, Ishibashi H (2004) Developmental switch from GABA to glycine release in single central synaptic terminals. Nat Neurosci 7:17–23

Nishimaki T, Jang IS, Ishibashi H, Yamaguchi J, Nabekura J (2007) Reduction of metabotropic glutamate receptor-mediated heterosynaptic inhibition of developing MNTB-LSO inhibitory synapses. Eur J NeuroSci 26:323–330

Norena AJ, Eggermont JJ (2003) Changes in spontaneous neural activity immediately after an acoustic trauma: implications for neural correlates of tinnitus. Hear Res 183:137–153

Norena AJ, Tomita M, Eggermont JJ (2003) Neural changes in cat auditory cortex after a transient pure-tone trauma. J Neurophysiol 90:2387–2401

Nusser Z, Cull-Candy S, Farrant M (1997) Differences in synaptic $GABA_A$ receptor number underlie variation in GABA mini amplitude. Neuron 19:697–709

Nusser Z, Hajos N, Somogyi P, Mody I (1998) Increased number of synaptic GABAA receptors underlies potentiation at hippocampal inhibitory synapses. Nature 395:172–177

Oda Y, Charpier S, Murayama Y, Suma C, Korn H (1995) Long-term potentiation of glycinergic inhibitory synaptic transmission. J Neurophysiol 74:1056–1074

Oda Y, Kawasaki K, Morita M, Korn H, Matsui H (1998) Inhibitory long-term potentiation underlies auditory conditioning of goldfish escape behavior. Nature 394(6689):182–185

Oleskevich S, Walmsley B (2002) Synaptic transmission in the auditory brainstem of normal and congenitally deaf mice. J Physiol 540:447–455

Oswald AM, Schiff ML, Reyes AD (2006) Synaptic mechanisms underlying auditory processing. Curr Opin Neurobiol 16:371–376

Owens DF, Boyce LH, Davis MB, Kriegstein AR (1996) Excitatory GABA responses in embryonic and neonatal cortical slices demonstrated by gramicidin perforated-patch recordings and calcium imaging. J Neurosci 16:6414–6423

Payne JA (1997) Functional characterization of the neuronal-specific K–Cl cotransporter: implications for $[K^+]o$ regulation. Am J Physiol 273:C1516–C1525

Payne JA, Stevenson TJ, Donaldson LF (1996) Molecular characterization of a putative K–Cl cotransporter in rat brain. J Biol Chem 27:16245–16252

Payne JA, Rivera C, Voipio J, Kaila K (2003) Cation-chloride co-transporters in neuronal communication, development, and trauma. Trends Neurosci 26:199–206

Paysan J, Kossel A, Bolz J, Fritschy JM (1997) Area-specific regulation of gamma-aminobutyric acid type A receptor subtypes by thalamic afferents in developing rat neocortex. Proc Natl Acad Sci 94:6995-7000.

Plotkin MD, Snyder EY, Hebert SC, Delpire E (1997) Expression of the Na–K–2Cl cotransporter is developmentally regulated in postnatal rat brains: a possible mechanism underlying GABA's excitatory role in immature brain. J Neurobiol 33:781–795

Pollak GD, Burger RM, Park TJ, Klug A, Bauer EE (2002) Roles of inhibition for transforming binaural properties in the brainstem auditory system. Hear Res 168:60–78

Pollak GD, Burger RM, Klug A (2003) Dissecting the circuitry of the auditory system. Trends Neurosci 26:33–39

Raggio MW, Schreiner CE (1999) Neuronal responses in cat primary auditory cortex to electrical cochlear stimulation. III. Activation patterns in short- and long-term deafness. J Neurophysiol 82:3506–3526

Raggio MW, Schreiner CE (2003) Neuronal responses in cat primary auditory cortex to electrical cochlear stimulation: IV. Activation pattern for sinusoidal stimulation. J Neurophysiol 89:3190–3204

Rajan R (1998) Receptor organ damage causes loss of cortical surround inhibition without topographic map plasticity. Nat Neurosci 1:138–143

Rajan R (2001) Plasticity of excitation and inhibition in the receptive field of primary auditory cortical neurons after limited receptor organ damage. Cereb Cortex 11:171–182

Rivera C, Voipio J, Payne JA, Ruusuvuori E, Lahtinen H, Lamsa K, Pirvola U, Saarma M, Kaila K (1999) The K$^+$/Cl$^-$ cotransporter KCC2 renders GABA hyperpolarizing during neuronal maturation. Nature 397:251–255

Romand R (1992) Development of Auditory and Vestibular Systems 2. Elsevier, Amsterdam

Russell FA, Moore DR (1995) Afferent reorganisation within the superior olivary complex of the gerbil: development and induction by neonatal, unilateral cochlear removal. J Comp Neurol 352:607–625

Salvi RJ, Wang J, Ding D (2000) Auditory plasticity and hyperactivity following cochlear damage. Hear Res 147:261–274

Sanes DH, Geary WA, Wooten GF, Rubel EW (1987) Quantitative distribution of the glycine receptor in the auditory brain stem of the gerbil. J Neurosci 7:3793–802

Sanes DH, Rubel EW (1988) The ontogeny of inhibition and excitation in the gerbil lateral superior olive. J Neurosci 8:682–700

Sanes DH, Siverls V (1991) Development and specificity of inhibitory terminal arborizations in the central nervous system. J Neurobiol 22:837–54

Sanes DH, Chokshi P (1992) Glycinergic transmission influences the development of dendritic shape. Neuro Report 3:323–326

Sanes DH, Markowitz S, Bernstein J, Wardlow J (1992) The influence of inhibitory afferents on the development of postsynaptic dendritic arbors. J Comp Neurol 321:637–644

Sanes DH (1993) The development of synaptic function and integration in the central auditory system. J Neurosci 13:2627–37

Sanes DH, Takács C (1993) Activity-dependent refinement of inhibitory connections. Eur J NeuroSci 5:570–574

Sanes DH, Harris WA, Reh TA (2006) Development of the Nervous System, 2nd edn. Academic, San Diego

Sarro E, Kotak VC, Sanes DH, Aoki C (2008) Hearing loss alters the subcellular distribution of presynaptic GAD and postsynaptic GABA$_A$ receptors in the auditory cortex. Cereb Cortex 18:2855–2867

Seki S, Eggermont JJ (2003) Changes in spontaneous firing rate and neural synchrony in cat primary auditory cortex after localized tone-induced hearing loss. Hear Res 180:28–38

Semple MN, Kitzes LM (1985) Single-unit responses in the inferior colliculus: different consequences of contralateral and ipsilateral auditory stimulation. J Neurophysiol 53:1467–1482

Shepherd RK, Baxi JH, Hardie NA (1999) Response of inferior colliculus neurons to electrical stimulation of the auditory nerve in neonatally deafened cats. J Neurophysiol 82:1363–1380

Snyder RL, Sinex DG, McGee JD, Walsh EW (2000) Acute spiral ganglion lesions change the tuning and tonotopic organziation of cat inferior colliculus neurons. Hear Res 147:200–220

Suneja SK, Potashner SJ, Benson CG (1998) Plastic changes in glycine and GABA release and uptake in adult brain stem auditory nuclei after unilateral middle ear ossicle removal and cochlear ablation. Exp Neurol 151:273–288

Syka J (2002) Plastic changes in the central auditory system after hearing loss, restoration of function, and during learning. Physiol Rev 82:601-636.

Szczepaniak WS, Moller AR (1995) Evidence of decreased GABAergic influence on temporal integration in the inferior colliculus following acute noise exposure: a study of evoked potentials in the rat. Neurosci Let 196:77–80

Takesian AE, Kotak VC, Sanes DH (2007) Sensorineural hearing loss disrupts Inhibitory short-term plasticity in the developing auditory cortex. Assoc Res Otolaryngol 30

Tan AY, Zhang LI, Merzenich MM, Schreiner CE (2004) Tone-evoked excitatory and inhibitory synaptic conductances of primary auditory cortex neurons. J Neurophysiol 92:630–643

Tang AH, Chai Z, Wang SQ (2007) Dark rearing alters the short-term synaptic plasticity in visual cortex. Neurosci Lett 422:49–53

Tehrani MH, Barnes EM Jr (1991) Agonist-dependent internalization of gamma-aminobutyric acid A/benzodiazepine receptors in chick cortical neurons. J Neurochem 57:1307–1312

Thornton S, Semple MN, Sanes DH (1999) Development of auditory motion processing in the gerbil inferior colliculus. Eur J NeuroSci 11:1414–1420

Tollin DJ (2003) The lateral superior olive: a functional role in sound source localization. Neuroscientist 9:127–143

Tritsch NX, Yi E, Gale JE, Glowatzki E, Bergles DE (2007) The origin of spontaneous activity in the developing auditory system. Nature 450:50–55

Tucci DL, Cant NB, Durham D (1999) Conductive hearing loss results in a decrease in central auditory system activity in the young gerbil. Laryngoscope 109:1359–1371

Tucci DL, Cant NB, Durham D (2001) Effects of conductive hearing loss on gerbil central auditory system activity in silence. Hear Res 155:124–132

Turrigiano G (2007) Homeostatic signaling: the positive side of negative feedback. Curr Opin Neurobiol 17:318–324

Vale C, Sanes DH (2000) Afferent regulation of inhibitory synaptic transmission in the developing auditory midbrain. J Neurosci 20:1912–1921

Vale C, Sanes DH (2002) The effect of bilateral deafness on excitatory and inhibitory synaptic strength in the inferior colliculus. Eur J NeuroSci 16:2394–2404

Vale C, Schoorlemmer J, Sanes DH (2003) Deafness disrupts chloride transporter function and inhibitory synaptic transmission. J Neurosci 23:7516–7524

Vale C, Juiz J, Moore D, Sanes DH (2004) Unilateral hearing loss produces greater loss of inhibition in the contralateral inferior colliculus. Eur J NeuroSci 20:2133–2140

Vale C, Caminos E, Martinez-Galán JR, Juiz JM (2005) Expression and developmental regulation of the K^+–Cl^- cotransporter KCC2 in the cochlear nucleus. Hear Res 206:107–115

Wang J, Ding D, Salvi RJ (2002a) Functional reorganization in chinchilla inferior colliculus associated with chronic and acute cochlear damage. Hear Res 168:238–249

Wang J, McFadden SL, Caspary D, Salvi R (2002b) Gamma-aminobutyric acid circuits shape response properties of auditory cortex neurons. Brain Res 19:219–231

Wang J, Reichling DB, Kyrozis A, MacDermott AB (1994) Developmental loss of GABA- and glycine-induced depolarization and Ca^{2+} transients in embryonic rat dorsal horn neurons in culture. Eur J NeuroSci 6:1275–1280

Wehr M, Zador AM (2003) Balanced inhibition underlies tuning and sharpens spike timing in auditory cortex. Nature 426:442–446

Wenthold RJ, Huie D, Altschuler RA, Reeks KA (1987) Glycine immunoreactivity localized in the cochlear nucleus and superior olivary complex. Neurosci 22:897–912

Wenthold RJ, Altschuler RA, Hampson DR (1990) Immunocytochemistry of neurotransmitter receptors. J Electron Microsc Tech 15:81–96

Werthat F, Alexandrova O, Grothe B, Koch U (2008) Experience-dependent refinement of the inhibitory axons projecting to the medial superior olive. Dev Neurobiol 68:1454–1462

Wiesel TN, Hubel DH (1965) Comparison of the effects of unilateral and bilateral eye closure on cortical unit responses in kittens. J Neurophysiol 28:1029–1040

Williams JR, Sharp JW, Kumari VG, Wilson M, Payne JA (1999) The neuron-specific K-Cl cotransporter, KCC2. Antibody development and initial characterization of the protein. J Biol Chem 274:12656–12664

Wisden W, Laurie DJ, Monyer H, Seeburg PH (1992) The Distribution of 13 GABA, Receptor Subunit mRNAs in the Rat Brain. I. Telencephalon, Diencephalon, Mesencephalon. J Neurosci 12:1040–1062

Woolf NK, Ryan AF (1984) The development of auditory function in the cochlea of the mongolian gerbil. Hear Res 13:277–283

Woolf NK, Ryan AF (1985) Ontogeny of neural discharge patterns in the ventral cochlear nucleus of the mongolian gerbil. Dev Brain Res 17:131–147

Wu SH, Kelly JB (1994) Physiological evidence for ipsilateral inhibition in the lateral superior olive: synaptic responses in mouse brain slice. Hear Res 73:57–64

Wyatt RM, Balice-Gordon RJ (2003) Activity-dependent elimination of neuromuscular synapses. J Neurocytol 32:777–794

Chapter 5
Developmental Plasticity of Inhibitory Receptive Field Properties in the Auditory and Visual Systems

Khaleel A. Razak, Zoltan M. Fuzessery, and Sarah L. Pallas

5.1 Introduction

Following the pioneering work of Hubel and Wiesel on developmental plasticity in the visual cortex, a considerable progress has been made in determining the relative contributions of experience-dependent and -independent mechanisms to the development of neural response properties (Katz and Shatz 1996, for review). During postnatal development in particular, neural activity driven by sensory inputs is critical for the refinement of response properties. Studies on the mechanisms, through which experience influences response selectivity, have focused on excitatory properties and connectivity. Only recently has the focus shifted toward inhibitory mechanisms (see Pallas et al, 2006; Huang et al. 2007; for recent reviews).

The appropriate balance between excitatory and inhibitory neural activity is critical for normal brain function. It is now well established that a number of response properties depend on the interactions between the inhibitory and excitatory portions of receptive fields (iRF and eRF). It is also known that inhibitory synapses show activity-dependent changes during development (Chattopadhyaya et al. 2004; Chen et al. 2001; Hensch and Fagiolini 2005; Kim and Kandler 2003; Morales et al. 2002; Turrigiano 1999; Vale et al. 2003). Whether such plasticity is adaptive or not depends on how changes in the inhibitory circuitry affect the response properties. However, few studies have addressed how activity-dependent plasticity of inhibitory synapses influences response selectivity (see e.g., Zheng and Knudsen 1999; Shoykhet et al., 2005; Razak and Pallas 2006 for supporting examples). This chapter

K.A. Razak (✉)
Department of Psychology & Graduate Neuroscience Program, University of California, 900 University Ave, Riverside, CA, 92521, USA

Z.M. Fuzessery
Department of Zoology and Physiology, University of Wyoming, Dept. 3166, 1000 E. University Ave, Laramie, WY, 82071, USA

S.L. Pallas
Neuroscience Institute, Georgia State University, 38 Peachtree Center Ave, Atlanta, GA, 30303, USA

summarizes our work on the role of inhibition in shaping selectivity for dynamic stimuli in the superior colliculus (SC) of hamsters and the auditory cortex (A1) of pallid bats, under normal and altered developmental conditions. The main conclusion is that the strength and timing of surround inhibition can be an important substrate upon which sensory experience acts to modify behaviorally relevant response selectivity.

5.1.1 Inhibitory Plasticity in the Hamster Superior Colliculus

The retinotectal/retinocollicular pathway of vertebrates has long been a model of choice for the studies of developmental plasticity (Sperry 1963; reviewed in Udin and Fawcett 1988; Constantine-Paton et al. 1990; Debski and Cline 2002). The superior colliculus (SC) is a midbrain structure involved in motion processing, and, as such, its retinorecipient neurons are selective for stimulus velocity (Rhoades and Chalupa 1978a; Stein and Dixon 1979; Razak et al. 2003). It is particularly rich in GABAergic terminals (Mize 1992, for review). The retinocollicular projection exhibits both activity-dependent and -independent forms of plasticity during development. We have shown that inhibition is important in velocity tuning in hamsters (Razak and Pallas 2005). Velocity-tuned neurons in the retinorecipient layers of hamster SC thus provide a suitable model for studying inhibitory plasticity of visual response properties. Here, we review findings on the role of inhibition in the development and plasticity of velocity tuning in the SC.

5.1.2 Surround Inhibition Shapes Velocity Tuning in the SC

Velocity tuning is a major characteristic of superficial SC (sSC) neurons, as expected in a structure involved in orienting attention to visual targets. Velocity tuning is remarkably resistant to developmental manipulations of activity and changes in afferent/target convergence ratio (Pallas and Finlay 1989; Huang and Pallas 2001; see below). Previous models of velocity tuning incorporated a directional component (Barlow et al. 1964). Because most sSC neurons are not directional except through cortical feedback (Rhoades and Chalupa 1978b), we undertook a study to uncover the circuitry underlying their velocity tuning.

An inhibitory region encircling the excitatory receptive field area is a common feature of visual system neurons. The majority of retinorecipient SC neurons exhibit such surround inhibition (Razak and Pallas 2005), meaning that the response to a stimulus in the RF is suppressed by a stimulus presented in the surround (Fig. 5.1a). Surround inhibition can be asymmetric (e.g., Fig. 5.1b) or symmetric (e.g., Fig. 5.1d) around the RF center. We found that the symmetry of the inhibitory surround and the type of velocity tuning in hamster SC are correlated, suggesting that there may be distinct classes of neurons as observed in cat SC (Waleszczyk, et al. 1999).

5 Developmental Plasticity of Inhibitory Receptive Field Properties in the Auditory

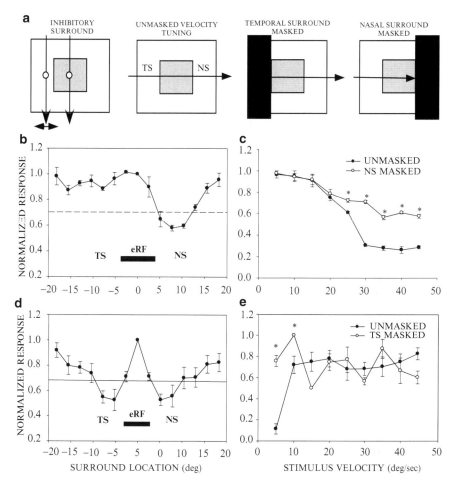

Fig. 5.1 Surround inhibition shapes velocity tuning in the SC. (**a**) Schematic of methodology used to determine surround inhibition and mechanisms of velocity tuning. The *gray square* represents the excitatory RF. The *white square* represents the surround. The surround was mapped using a two-spot stimulus. One spot was swept vertically through the center of the RF, as a second, simultaneous spot was swept at progressively greater distances from the first at 2.6° increments. *TS* temporal surround, *NS* nasal surround. Velocity tuning was determined using a stimulus moving in a temporal to nasal direction at velocities between 5 and 45°. The contribution of surround inhibition to velocity tuning was determined by masking different parts of the inhibitory surround. (**b**) A typical neuron with asymmetric surround. The black rectangle denotes the extent of the eRF. This neuron exhibited inhibition on the NS, but not on the TS. (**c**) The NS contributes to velocity tuning in this LP neuron because masking the nasal NS reduces velocity selectivity. In most LP neurons blocking the NS, but not the TS reduces velocity tuning. (**d**) A typical neuron with symmetric surround inhibition. (**e**) In this HP neuron, masking the TS virtually eliminates selectivity. In most HP neurons, blocking the TS, but not the NS reduces velocity tuning. * $p<0.05$. Figure adapted from Razak and Pallas (2005)

All neurons with asymmetric surround inhibition were selective for slowly moving stimuli (low-pass-LP neurons, e.g., Fig. 5.1c). LP tuning can be accounted for, at least in part, by a form of temporal asymmetry called backward masking, i.e. inhibition arising from the surround traversed after the stimulus leaves the RF and suppresses responses to rapidly moving stimuli (Fig. 5.1c). On the other hand, most SC neurons with symmetric surrounds (Fig. 5.1d) prefer rapidly moving stimuli (high-pass-HP neurons). Despite the symmetry, only the surround location traversed before the stimulus enters the RF (forward masking) contributes to HP tuning, by reducing responses to slowly moving stimuli before they enter the eRF (Fig. 5.1e)

The masking data, while supporting the importance of inhibition in creating velocity tuning, did not reveal where that inhibition is located. To address this issue, we applied GABA-A receptor antagonists iontophoretically on SC neurons during electrophysiological recordings and found that velocity tuning is shaped by intra-SC GABA in nearly half the population (Khoryevin, Razak and Pallas, in preparation). These data suggest that the surround inhibition shaping velocity tuning arises, at least partly, from neurons intrinsic to the SC.

Taken together, these data suggest that the spatiotemporal interactions between the inhibitory surround and the eRF shape the velocity tuning in the SC. The amount of time a moving stimulus spends in the different spatial components of the visual field (NS, RF, TS, see Fig. 5.1) depends on both the size of the components and the velocity of the movement. For a given stimulus velocity, it can be predicted that the timing of excitatory and inhibitory inputs triggered by a moving stimulus will depend on the spatial extent of the RF components. Thus, we expected that velocity tuning would be altered by changes in the RF size.

5.1.3 *Effects of Modifying Retinocollicular Convergence on Surround Inhibition During Development*

Retinocollicular convergence ratios decrease during normal development in hamsters (Schneider 1973; Huang and Pallas 2001). Hamsters open their eyes at approximately P12, and at this age, RFs are large and diffuse. Between P25 and P50, the average eRF diameter of SC neurons becomes smaller, revealing a postnatal refinement process (Carrasco et al. 2005). This refinement is NMDA receptor (NMDAR)-activity dependent in rodents (Simon et al., 1992; Huang and Pallas 2001; Colonnese and Constantine-Paton 2006), although visual experience is not necessary (Carrasco et al. 2005), suggesting that spontaneous glutamatergic activity is the critical factor. Chronic postnatal blockade of NMDAR using the selective antagonist D-APV results in increased RF diameters of 50%, on average (Huang and Pallas 2001). Thus, a light spot moving through the RF will spend more time in the RF in D-APV-exposed neurons than in normal neurons. Contrary to our prediction that this alteration in the time course over which the inhibitory and excitatory inputs interact would result in altered velocity tuning, however, the velocity tuning showed no difference between the normal and the D-APV groups (Razak et al. 2003). How can this be explained?

We hypothesized that the spatiotemporal relationships underlying velocity tuning were maintained through concomitant changes in surround inhibition, following an increase in the RF size (Razak and Pallas 2007). This predicts first that the surround inhibition increases in spatial extent, and secondly, that it makes a larger contribution to velocity tuning in the D-APV group than in the normal group. We tested the first prediction by comparing the surround inhibition between normal and D-APV groups. We tested the second prediction by comparing the percentage reduction in velocity tuning following masking of surround inhibition between the two groups. We found that chronic postnatal NMDAR blockade increased the strength of surround inhibition in SC neurons (schematized in Fig. 5.2a, b). The size of the RF of individual neurons was correlated with the size of the inhibitory surround in the D-APV group, suggesting a matching change in both excitatory and inhibitory RF regions, following NMDAR blockade.

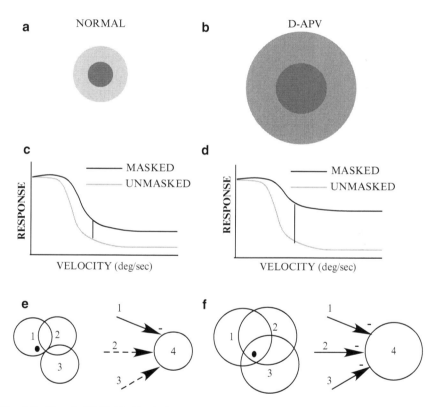

Fig. 5.2 Chronic NMDAR blockade increases the spatial extent and strength of surround inhibition in the SC. (**a, b**) Schematic illustration showing that both excitatory RF and surround increase in size following chronic NMDAR blockade. The strength of inhibition also increases in the D-APV group (indicated by the *darker shade of gray*). (**c, d**) There was a larger increase in response to non-optimal velocities following masking of the surround in the D-APV group compared to normal. (**e, f**) Possible mechanism underlying increased strength and spatial extent of surround inhibition. See text for details. Figure adapted from Razak et al. (2003) and Razak and Pallas (2007)

5.1.4 Surround Inhibition Plays a Larger Role in Velocity Tuning After Chronic NMDAR Blockade

Similar to normal SC neurons, in the D-APV group backward masking shapes LP and forward masking shapes HP velocity tuning. However, masking the surround resulted in a greater reduction of velocity tuning in the D-APV group compared to the normal group in both LP and HP neurons (Fig. 5.2c, d). These results suggest that the increased strength and extent of surround inhibition in the D-APV group contributes to the maintenance of LP/HP velocity tuning following experimental increases in eRF diameter. Thus, it appears that NMDAR activity plays an indirect role in shaping the velocity tuning in the SC by refining the spatial extents of the interacting excitatory and inhibitory RF components (Fig. 5.2e, f). This counterbalancing change results in the maintenance of velocity tuning despite variation in the extent of inputs resulting from manipulations of activity. Thus, we see that when the spatial extent of excitation is increased, inhibitory plasticity preserves the function of the circuit, presumably allowing the animal to continue making appropriate visual discriminations of the moving targets. Not only is this process in operation after experimental manipulations, but is, likely, also important in the normal refinement of receptive fields during development.

5.1.5 Plasticity of Inhibition Underlying Vocalization Selectivity in the Auditory Cortex

Frequency-modulated (FM) sweeps are analogous to moving visual stimuli in that both classes of stimuli contain movements across the sensory epithelium, allowing a comparison of how different sensory modalities solve analogous problems. FM sweeps are common in species-specific vocalizations, including human speech. Abnormalities in FM sweep processing may underlie the deficits in speech processing (Merzenich et al. 1996). Neurons selective for FM sweep direction and rate have been found in every species examined (Suga 1969; Heil et al. 1992; Mendelson et al. 1993; Nelken and Versnel 2000; Tian and Rauschecker 2004), but the role of experience in the development of FM sweep selectivity has not been examined. An understanding of how FM sweeps are represented and how such representation develops will provide important insights into the development of vocalization representation in general.

The pallid bat is suited to address this issue due to the relatively simple FM sweep (downward sweep, 60–30 kHz, 2-5 ms duration) it uses to echolocate, and the strong selectivity in the auditory system for the downward direction and a narrow range of FM sweep rates (See Fig. 5.3a, for an example). We studied the role of inhibition in shaping the selectivity for FM sweeps and how such inhibitory mechanisms are modified by developmental experience. Relevant to this section are

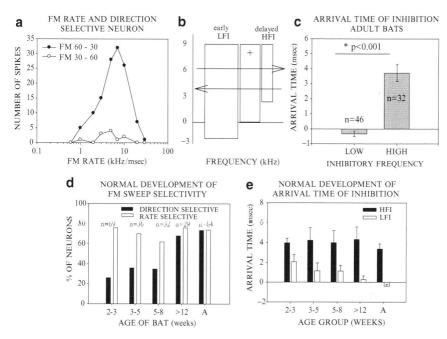

Fig. 5.3 Mechanisms and development of FM rate and direction-selectivity in the pallid bat auditory cortex. (**a**) A typical FM rate and direction selective neuron. A neuron was classified as rate selective if the response fell to less than 50% of the maximum response for decreasing FM rates. Direction selectivity was determined by comparing responses to FM sweeps of the same bandwidth, but sweeping in opposite directions. A neuron was considered to be direction selective if the maximum response to one-sweep direction was lower than 25% of the maximum response to the opposite-sweep direction. The neuron shown here clearly responded better to the 60–30 kHz downward sweep, than to the 30–60 kHz upward sweep. (**b**) A two-tone inhibition over time (TTI) plot. The arrival time of inhibition was determined by presenting two tones with various delays between them. One tone was excitatory, while the other tone was inhibitory. The delay of the excitatory tone relative to the inhibitory tone is shown on the y-axis, with positive (negative) delays indicating that the excitatory tone was delayed (advanced). In the sample TTI plot shown, low-frequency inhibition (LFI) arrives early. High-frequency inhibition (HFI) is delayed. (**c**) In the pallid bat auditory cortex, LFI arrived early and HFI was delayed in the majority of neurons. (**d**) During normal development, the percentage of rate-selective neurons is adult-like from P14 onwards. Direction selectivity is slow to mature, with only 25% of neurons exhibiting direction selectivity at P14. An adult-like percentage of direction-selective neurons is observed after P90. (**e**) Development of arrival times of inhibition. HFI is adult-like from P14, and underlies adult-like rate selectivity from P14. LFI arrival time is delayed in pups. On average, LFI arrives progressively earlier with age, matching the development of adult-like direction selectivity. Figure adapted from Razak and Fuzessery (2006) and Razak and Fuzessery (2007)

the findings that FM sweep selectivity is shaped by inhibitory RF properties, and these inhibitory properties are strongly influenced by developmental experience. An important finding is that experience is required for the maintenance of innately specified response properties.

5.1.6 Asymmetries in Sideband Inhibition Shape FM Rate and Direction Selectivity in Adults

Auditory cortical neurons are similar to other sensory cortical neurons in that they have a center-surround excitatory–inhibitory RF organization. Suga (1969) first proposed that asymmetries in surround (sideband) inhibition underlie FM sweep selectivity. In addition to the presence or absence of a sideband inhibition, a key determinant of sweep selectivity is the relative arrival time of inhibition (Razak and Fuzessery 2006). If excitatory and inhibitory tones are presented simultaneously, the ipsp may occur before, together with, or after the epsp, depending on the effective arrival time of each input. The arrival time of inhibition can be inferred by presenting tone frequencies that evoke inhibition or excitation at different delays with respect to each other. If a response suppression occurs even if the inhibitory tone is delayed relative to the excitatory tone (backward masking), then it can be inferred that the inhibition arrived early. If it occurs only when the inhibitory tone is presented before the excitatory tone (forward masking), then the inhibition is delayed.

Consider the hypothetical RF structure shown in Fig. 5.3b. In this neuron, low-frequency inhibition (LFI) arrives earlier than excitation by 1 ms (shown as starting at negative delays), whereas high-frequency inhibition (HFI) arrives 3 ms later than excitation (shown as arriving only at positive delays). During an upward FM sweep that includes the frequencies evoking LFI, inhibition will be generated first, suppressing excitation. For a downward FM sweep that includes frequencies evoking the HFI, the delayed HFI allows the neuron to respond to downward sweeps, if the sweep is fast enough and reaches the eRF before inhibition can catch up. If the sweep is slow, the HFI will arrive first and will suppress responses. The result is selectivity for the rate of downward FM sweeps. Thus, early LFI and late HFI can theoretically generate direction and rate selectivity, respectively, for downward sweeps.

In adult pallid bat auditory cortex, where neurons tuned in the echolocation range of frequencies, LFI arrives early, whereas HFI is delayed (Fig. 5.3c, Razak and Fuzessery 2006). This gives rise to direction and rate selectivity according to the model presented in Fig. 5.3b. Direction selectivity in these neurons was reduced when the LFI was excluded from the upward sweep by starting the sweep inside the RF (analgous to visual RF masking experiments). Rate-selectivity for downward sweeps is eliminated when the downward FM sweep excludes the HFI. Neurons without HFI were not rate selective for downward sweeps. Taken together, these data show that FM rate and direction selectivity are shaped by temporal asymmetries in the sideband inhibition. Thus, the question of how direction and rate selectivity mature during normal development can be reformulated as how the timing of LFI and HFI change during a development.

5.1.7 Developmental Plasticity of Inhibition Underlying FM Rate and Direction Selectivity

Pallid bats begin to develop hearing sensitivity to frequencies used in echolocation after P11 (Brown 1976). We found that FM rate selectivity (Fig. 5.3d) and the underlying HFI arrival time (Fig. 5.3e) were similar between P14 pups and adults (Razak and Fuzessery 2007). Because rate selectivity is adult-like at the time when the bat's audiogram is first adult-like (~P14), we conclude that rate selectivity develops in an experience-independent manner. Direction selectivity, however, was present only in ~25% of neurons at P14 (Fig. 5.3d). Direction selectivity and the underlying mechanism (LFI arrival time) become adult-like after 12 weeks (Fig. 5.3d, e). These data show that the adult-like complement of direction-selective neurons develops well after the onset of hearing in the echolocation range and arises through a developmental advancement of LFI arrival time. Thus, it appears that the pallid bat is born with an innate selectivity for the rate of change of frequencies present in the adult echolocation call. A part of this template is the delayed HFI. Direction selectivity and the underlying LFI arrival time develop slowly, and may be shaped by experience.

5.1.8 Experience-Dependent Plasticity of Inhibition Shaping Rate and Direction Selectivity

To test the role of experience in the development of FM rate and direction selectivity, we eliminated normal experience with echolocation calls during development (Razak et al. 2008). Pups were muted before P13 either by lesioning the laryngeal muscles with heat or by injecting botulinum toxin A (Botox) into the laryngeal muscles. The muted pups were acoustically and physically isolated from other bats. We compared rate and direction selectivity and the underlying inhibitory mechanisms between the normal and muted pups at P30 and P90. To control for the isolation in the muted group, we also included a group of pups that were isolated, but not muted. These control pups also served to determine if self-vocalizations were sufficient to generate normal calls and response properties.

Laryngeal manipulations altered, but did not eliminate, the production of echolocation calls. During development, normal and control pups produced adult-like calls from P20 onward. However, the muted pups produced calls with significantly lower frequencies and rate of change of frequencies, at all ages up to P90. Thus, muted pups were deprived of normal experience with echolocation calls, until the day of electrophysiological recordings, allowing us to ask how altered experience influenced response selectivity for echolocation calls and timing of inhibition.

5.1.9 Normal Experience is Required for the Maintenance of FM Rate Selectivity and HFI

The muted pups showed a significantly lower percentage of FM rate selective neurons compared to age-matched control and normal pups (Fig. 5.4a). Because rate selectivity is adult-like in P14 normal pups, the muted group data suggest that

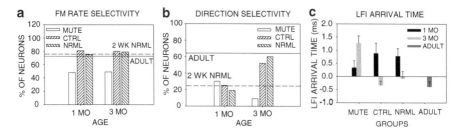

Fig. 5.4 Experience-dependent plasticity of inhibitory mechanisms underlying FM sweep selectivity. (**a**) The percentage of rate-selective neurons was significantly reduced in the muted pups compared with normal and control pups at P30 and P90. The percentage was also lower compared to P14 normal pups (*dashed horizontal line*) and adults (*solid horizontal line*). (**b**) Direction selectivity was similar across the three groups at P30. A dramatic decrease in the percentage of direction-selective neurons was observed in the P90-muted group. (**c**) In the normal and control groups, LFI arrives significantly earlier at P90 compared to P30. However, in the mute group, LFI arrival time is delayed at P90 compared to P30. Figure adapted from Razak et al. (2008)

normal experience is required not for the initial development of FM rate selectivity, but for its maintenance. A higher percentage of neurons in the muted group lack HFI compared to neurons in the normal and control groups (shown as HFI absent in Fig. 5.5). Thus, normal experience is required for the maintenance of HFI underlying FM rate selectivity.

5.1.10 Experience is Required for Development and Maintenance of Direction Selectivity and LFI

At P30, the percentage of direction-selective neurons in the muted pups was similar to that observed in the normal and control pups (Fig. 5.4b). However, a dramatic decrease in the percentage of direction-selective neurons was observed in the muted group at P90 (Fig. 5.4c), resulting from both a failure of complete development and a loss of direction selectivity compared to initial levels. These data show that both development and maintenance of direction selectivity requires normal experience with echolocation calls.

The reduction in direction selectivity in muted pups was either due to a loss of LFI (shown as absent LFI in Fig. 5.5) or a delay in its arrival time (Fig. 5.4c and Fig. 5.5). A significantly higher percentage of neurons in the P90-muted pups exhibited either delayed LFI or lacked LFI, altogether when compared to control and normal pups. Because an early-arriving LFI is critical for direction selectivity, these data show that experience-dependent changes of timing of inhibition in the millisecond range can have a significant impact on refinement and maintenance of response selectivity.

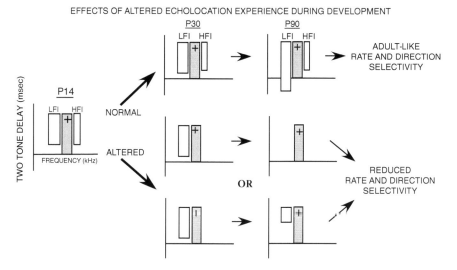

Fig. 5.5 Normal echolocation experience is required for the development of inhibitory mechanisms underlying FM sweep selectivity. Electrophysiological recordings from normal P14–P90 pups show that the timing of HFI is adult-like from p14 resulting in adult-like FM-rate selectivity from the time the pups first hear echolocation frequencies. LFI, however, is on average delayed at P14 and P30 compared to adults. This results in a lower incidence of direction-selective neurons. Only at P90 does LFI timing and direction selectivity become adult-like. In pups developing without normal echolocation experience, LFI timing and direction selectivity are similar to the normal pups at P30. HFI and rate selectivity are absent in a larger percentage of neurons in P30 mutes. At P90, LFI is either absent or delayed in a greater percentage of neurons in the muted pups. These data show that rate and direction selectivity are shaped by echolocation experience through modification of sideband inhibition

5.2 Discussion

5.2.1 *The Contribution of Surround Inhibition to RF Properties Across Sensory Systems*

Surround (or sideband) inhibition is important for direction and velocity (rate) selectivity in the visual and auditory systems. Both systems exhibit asymmetries in surround inhibition, indicating similar solutions to the construction of spatio/spectro-temporal filters (Razak and Fuzessery 2008). In the SC, nearly a third of the neurons exhibited spatial asymmetries in the surround, with stronger inhibition on the nasal than the temporal side of the RF. Temporal asymmetries can also be inferred based on the backward/forward masking data. In the auditory cortex, there was an asymmetry in timing, with LFI arriving earlier than HFI, relative to excitation. Spectral bandwidth (the cochlear analogue to retinal space) was also asymmetric with LFI being broader than HFI (data not shown, but see Razak and Fuzessery 2006).

The presence of spatially asymmetric surround inhibition has also been reported in the visual cortex of cats (Walker et al. 1999), although the role of this asymmetry remains unclear. In the auditory cortex and SC, asymmetries in surround inhibition shape direction and velocity (rate) selectivity, as evidenced by the reduction or loss of selectivity if the influence from specific surround locations/frequencies is removed.

In both SC and auditory cortex, surround inhibition is plastic during development. In the auditory cortex, the data taken between P14 and adulthood show that the temporal asymmetries in the arrival time of HFI and LFI become more pronounced with age and experience, particularly due to changes in the arrival time of LFI. LFI starts out delayed and advances systematically throughout development. Bats without exposure to normal echolocation calls show either a loss of or a delay in the arrival time of sideband inhibition. In the SC, surround inhibition is altered in its strength and spatial extent by activity dependent changes in the size of the eRF, presumably through increased retinocollicular convergence ratios. It is unknown if there is a change in the timing of inhibition from the surround as well. In both the auditory and the visual systems, the change in surround inhibition leads to measurable effects on neural selectivity to dynamic properties of stimuli. Thus, the changes in surround inhibition may be a common substrate for experience-dependent plasticity.

5.2.2 Previous Studies on the Role of Inhibitory Plasticity in the Development of Response Selectivity

A role for inhibition in adaptive plasticity during development was first shown by Zheng and Knudsen (1999) based on their work in prism-reared barn owls. Auditory space-tuned neurons in the external nucleus of the inferior colliculus project to the optic tectum, where maps of auditory and visual space are arranged in spatial register. Space-tuned neurons depend, in part, on sensitivity to interaural time differences (ITD) for azimuth tuning. The layout of the auditory space map in the tectum is under the direction of the visual map (reviewed in Kundsen 2002). Altering the visual map during development by raising owls with prisms over their eyes causes an adaptive shift in the auditory map to follow the movement of the visual map. Zheng and Knudsen (1999) showed that the ITD sensitivity of auditory neurons in the ICx of owls shifts in an adaptive direction dictated by prism-induced changes in visual locations. Application of GABA-A receptor antagonists unmasks ITD sensitivity corresponding to pre-prism spatial locations, suggesting that the original ITD sensitivity was masked by the plasticity of inhibitory synapses. Thus, inhibitory plasticity underlies the adaptive change.

The role of experience in shaping RF structure through inhibitory plasticity has been tested in the somatosensory, visual, and auditory systems. In rat somatosensory cortex, whisker trimming during early stages of development causes a reduction in suppressive interactions in adults after whisker regrowth, suggesting that early experience shapes the inhibitory–excitatory balance necessary for RF refinement (Shoyket et al., 2005; Sun, this volume). In the hamster SC, dark-rearing reduces surround inhibition, leading to an increase in the RF size (Carrasco et al. 2005, 2009). There is also a reduction of inhibition inside the RF, leading to a broadening of stimulus size tuning (Razak and Pallas 2006). In the auditory cortex of rats, rearing in a continuous, moderately noisy environment caused a disruption in the maturation of both spectral and temporal properties of the inhibitory surround (Chang et al. 2005). Taken together with our results, these data suggest that inhibitory plasticity underlies activity-dependent plasticity, resulting in both adaptive and abnormal changes in RF properties.

5.2.3 Homeostatic Plasticity of Inhibition: Beyond Response Magnitude Stability

Based on the studies of the visual system, it has been proposed that the development of response properties progresses in two stages (reviewed in Constantine-Paton et al. 1990) The initial establishment of underlying circuits is experience-independent. The second stage involves experience-dependent refinement of these circuits. Plasticity of inhibition during the period of refinement has primarily been discussed in terms of homeostatic balance of response magnitude (Turrigiano and Nelson 2004; Akerman and Cline 2007). Based on this view, inhibitory synaptic strength is altered to match activity-dependent changes in excitatory inputs or intrinsic excitability (Karmarkar and Buonomano 2006). For example, visual deprivation causes a decrease in the excitatory synaptic drive from the retina, resulting in a decrease in the inhibitory drive and thus maintaining stable levels of activity in the developing visual cortex (Turrigiano 1999; Maffei et al. 2004, 2006). A similar homeostatic shift to maintain the balance between excitation and inhibition occurs in the developing auditory cortex, following hearing loss (Kotak et al. 2005; this volume) and in the neuromuscular system of activity-deprived chick embryos (Gonzalez-Islas and Wenner 2006; this volume).

Our results extend these findings by showing that factors in addition to stability of response magnitude drive the plasticity of inhibition. In the auditory cortex, development involves appropriate matching of arrival times of inhibition and excitation. In the SC, the size of the inhibitory surround changes with the excitatory RF size. Thus, inhibitory plasticity may also function in balancing the timing and spatial/spectral extents of excitatory and inhibitory inputs. These results further suggest that the response magnitude is not always the conserved commodity, when homeostatic plasticity occurs following developmental manipulation of activity. In hamster SC, light-evoked activity is primarily due to AMPA receptor currents, and chronic NMDAR blockade does not significantly alter the levels of glutamate-evoked or light-evoked activity (Huang and Pallas 2001). Therefore, the animals reared with chronic NMDAR blockade may not experience reductions in excitatory SC activity during development; yet, we observed an increase in the strength of surround inhibition. The comparison of results from the SC of NMDAR-blocked (Razak and Pallas 2007) and dark-reared (Carrasco et al. 2005, 2006, and submitted) hamsters shows different directions of plasticity of surround inhibition. The former shows an increase in surround inhibition, while the latter shows a decrease, illustrating the complexity of factors shaping the development of the balance between inhibition and excitation.

5.2.4 Possible Synaptic Mechanisms of Plasticity in Strength and Timing of Inhibition

In the SC, although it is possible that NMDAR blockade directly increases the effectiveness of GABAergic synapses (Shi et al. 1997), an alternative explanation for our results may be that the increase in the strength and spatial extent of the inhibitory surround is an indirect effect of chronic NMDAR blockade. In rodent SC, GABAergic interneurons are themselves likely to have a larger eRF as a result of the D-APV-induced increase in the retinal convergence. A visual stimulus would, thus, excite more inhibitory neurons in the D-APV group than normal, and could result in the observed increase in the strength and extent of the inhibitory surround (Fig. 5.2e, f). This compensation mechanism could function during development or evolution, or as a mechanism for recovery from abnormal experience or trauma in sensory circuits, in general.

The reduction in the strength of sideband inhibition in the auditory cortex of bats raised with abnormal echolocation experience may result from reducing the strength of synaptic inhibition, possibly through a loss of $GABA_A$ receptors from the synapse (Kilman et al. 2002), modulation of chloride transporter function (Vale et al. 2003), phosphorylation of $GABA_A$ receptors (reviewed in Mody 2005), changes in subunit composition of $GABA_A$ receptors (Ortinski et al. 2004), changes in the number of synaptic vesicles (Murthy et al. 2001), or other forms of presynaptic modulation of GABA release (Morales et al. 2002; Misgeld et al. 2007). Future studies will attempt to determine the cellular mechanisms underlying altered strength of surround inhibition.

Abnormal echolocation experience also results in the delay in the arrival time of LFI. How the timing of inhibition is altered by experience remains unclear. One possibility is that the strength and timing of inhibitory inputs are related. That is, a delay in arrival time may be caused by a reduction in the strength of the inhibitory input. Wu et al., (2006) showed that in the rat auditory cortex, the timing of inhibitory currents varied monotonically with the intensity of tones. Preliminary data from the Razak lab show that increasing the intensity of the inhibitory tones, with respect to the excitatory tones in the two-tone inhibition protocol, can cause significant changes in the arrival times. Thus, a critical developmental event across sensory systems may be the matching of amplitudes of inhibitory and excitatory inputs.

Another possible mechanism for changes in the timing of inhibition and its effect on FM direction selectivity is spike-timing dependent plasticity (STDP). During the development of the pallid bat auditory system, the dominant-patterned input to neurons involved in echolocation, is likely to be downward FM sweeps. Consistent exposure to downward sweeps, with a small range of FM rates (rate of change of frequencies in the sweep), may cause neurons to favor inputs that are coactivated with the spectrotemporal relationships of inhibition and excitation naturally present in those sweeps (Engert et al. 2002). The slower FM sweep rates and the reduced high frequencies that the muted pups experience have a different spectrotemporal sequence compared to normal echolocation calls, and the coincident pre-synaptic events may not be driven by the same combination of inputs that drive neurons in the normal group. This would result in weakening of synapses established in an experience-independent manner and/or prevention of experience-dependent refinement (for detail, see Razak and Fuzessery 2007).

5.2.5 *Role of Experience During Development: Maintenance Versus Refinement*

Experience is typically thought to play an important role in the refinement of neural response selectivity in sensory systems. Our data suggest that experience can also be important for maintaining response selectivity that was originally created in an experience-independent manner. Few other studies have looked at the role of experience in the maintenance of response properties. In ferrets, blocking retinal activity after eye-specific segregation has occurred in the lateral geniculate nucleus causes desegregation (Chapman 2000), suggesting that activity is required for the maintenance of connectivity. We have shown that in the SC of hamsters, receptive fields refine in the absence of light during development (Carrasco et al. 2005). However, continued maintenance of the animals in the dark causes RF diameters to broaden, suggesting that light input is required for the maintenance of RF size. The reduction in surround inhibition in hamsters maintained in the dark is suggestive of a role for inhibitory mechanisms; and indeed, the blunted response to GABA agonist and antagonists in the dark-reared animals supports this interpretation (Carrasco and Pallas 2007, submitted).

Therefore, data from the visual and auditory systems together suggest that experience plays a key role in maintaining response properties in sensory systems, and often acts through modifications of inhibitory properties. One implication of these findings is that previous studies on the effects of sensory deprivation during early development on response properties in adults may have confused effects on refinement with effects on maintenance.

5.2.6 Future Directions

The hamster SC and pallid bat auditory cortex are suitable models to study the plasticity of inhibitory RF properties within a behaviorally relevant context. The findings reported here raise several key questions that need to be addressed in the future:
1. Although the presence of surround inhibition is common in visual system neurons, the development of spatiotemporal properties of surround inhibition and the role of corticofugal connections have not been widely studied.
2. To determine if the timing of inhibition is related to inhibitory synaptic strength, and why LFI is specifically delayed by altered echolocation experience, *in vivo* intracellular recordings are needed to elucidate the inhibitory and excitatory inputs at various sound frequencies in normal and muted pallid bats. These studies will provide clues about the synaptic mechanisms of response selectivity and plasticity in the auditory cortex.
3. The origin of plasticity observed in the auditory cortex is unclear. FM sweep selectivity is similar in the auditory cortex and inferior colliculus (IC) of the pallid bat. Whether changes in the cortex are inherited from the IC, or whether cortical changes influence the IC is not known.
4. An issue of considerable interest is the role of inhibition in establishing the critical period for experience-dependent circuit refinement. A certain threshold of tonic inhibition is required to trigger the onset of critical periods, and the onset can be advanced or postponed by manipulations of inhibition (Huang et al. 1999; Iwai et al. 2003). It must be noted that we have studied the development of phasic (stimulus driven) inhibition. It remains unclear how tonic and phasic inhibition interact during development to establish multiple critical-period windows.
5. Whether critical periods for plasticity of excitatory and inhibitory mechanisms are similar remains unclear.
6. Perhaps the most fundamental question about inhibitory plasticity is the underlying mechanism. While Hebbian and STDP-based mechanisms predict plasticity at excitatory synapses, it remains unclear whether they can explain plasticity at inhibitory synapses (but see Woodin et al. 2003 and Nugent et al. 2007). The synaptic mechanisms of inhibitory plasticity need to be addressed.

Acknowledgments We thank the members of the Pallas and Fuzessery labs for commenting on an earlier version of this chapter. Funding was provided by NIDCD DC05202 to ZMF, NIH EY12696, NSF IBN-0078110 and Georgia State University Research Foundation grants to SLP and National Organization for Hearing Research Foundation grant to KAR.

References

Akerman CJ, Cline HT (2007) Refining the roles of GABAergic signaling during neural circuit formation. Trends Neurosci 30:382–389

Barlow HB, Hill RM, Levick WR (1964) Retinal ganglion cells responding selectively to direction and speed of image motion in the rabbit. J Physiol 173:377–407

Brown P (1976) Vocal communication in the pallid bat, *Antrozous pallidus*. Z Tierpsychol 41(34–54):1976

Carrasco MM, Razak KA, Pallas SL (2005) Visual experience is necessary for maintenance but not development of receptive fields in superior colliculus. J Neurophysiol 94:1962–1970

Carrasco MM, Pallas SL (2006) Early visual experience prevents but cannot reverse deprivation-induced loss of refinement in adult superior colliculus. Vis Neurosci 23:845–852

Carrasco MM, Mao Y-T, Pallas SL (2007) Adult plasticity in superior colliculus of dark-reared hamsters results from loss of surround inhibition. Neuroscience Meeting Planner. San Diego, CA: Society for Neuroscience, 2007. Online Program No 36.10.

Chang EF, Bao S, Imaizumi K, Schreiner C, Merzenich MM (2005) Development of spectral and temporal response selectivity in the auditory cortex. Proc Natl Acad Sci USA 102:16460–16465

Chapman B (2000) Necessity for afferent activity to maintain eye-specific segregation in ferret lateral geniculate nucleus. Science 287:2479–2482

Chattopadhyaya B, Di Cristo G, Higashiyama H, Knott GW, Kuhlman SJ, Welker E, Huang ZJ (2004) Experience and activity-dependent maturation of perisomatic GABAergic innervation in primary visual cortex during a postnatal critical period. J Neurosci 24:9598–611

Chen L, Yang C, Mower GD (2001) Developmental changes in the expression of GABA(A) receptor subunits (alpha(1), alpha(2), alpha(3)) in the cat visual cortex and the effects of dark rearing. Brain Res Mol Brain Res 88(1–2):135–143

Colonnese MT, Constantine-Paton M (2006) Developmental period for N-methyl-D-aspartate (NMDA) receptor-dependent synapse elimination correlated with visuotopic map refinement. J Comp Neurol 494:738–751

Constantine-Paton M, Cline HT, Debski E (1990) Patterned activity, synaptic convergence, and the NMDA receptor in developing visual pathways. Annu Rev Neurosci 13:129–154

Debski EA, Cline HT (2002) Activity-dependent mapping in the retinotectal projection. Curr Opin Neurobiol 12:93–99

Engert F, Tao HW, Zhang LI, Poo MM (2002) Moving visual stimuli rapidly induce direction sensitivity of developing tectal neurons. Nature 419:470–475

Finlay BL, Schneps SE, Schneider GE (1979) Orderly compression of the retinotectal projection following partial tectal ablation in the newborn hamster. Nature 280(5718):153–155

Gonzalez-Islas C, Wenner P (2006) Spontaneous network activity in the embryonic spinal cord regulates AMPAergic and GABAergic synaptic strength. Neuron 49:563–575

Heil P, Rajan R, Irvine DRF (1992) Sensitivity of neurons in cat primary auditory cortex to tones and frequency modulated stimuli. II.Organization of response properties along the 'isofrequency' dimension. Hear Res 63:135–156

Hensch TK (2004) Critical period regulation. Annu Rev Neurosci 27:549–579

Hensch TK, Fagiolini M (2005) Excitatory-inhibitory balance and critical period plasticity in developing visual cortex. Prog Brain Res 147:115–124

Huang L, Pallas SL (2001) NMDA antagonists in the superior colliculus prevent developmental plasticity but not visual transmission or map compression. J Neurophysiol 86:1179–1194

Huang ZJ, Kirkwood A, Pizzorusso T, Porciatti V, Morales B, Bear MF, Maffei L, Tonegawa S (1999) BDNF regulates the maturation of inhibition and the critical period of plasticity in mouse visual cortex. Cell 98:739–755

Huang ZJ, Di Cristo G, Ango F (2007) Development of GABA innervation in the cerebral and cerebellar cortices. Nat Rev Neurosci 8:673–686

Iwai Y, Fagiolini M, Obata K, Hensch TK (2003) Rapid critical period induction by tonic inhibition in visual cortex. J Neurosci 23(17):6695–6702

Karmarkar UR, Buonomano DV (2006) Different forms of homeostatic plasticity are engaged with distinct temporal profiles. Eur J NeuroSci 23:1575–1584

Katz LC, Shatz CJ (1996) Synaptic activity and the construction of cortical circuits. Science 274:1133–1138

Katz LC, Crowley JC (2002) Development of cortical circuits: lessons from ocular dominance columns. Nat Rev Neurosci 3:34–42

Kilman V, van Rossum MC, Turrigiano GG (2002) Activity deprivation reduces miniature IPSC amplitude by decreasing the number of postsynaptic GABA(A) receptors clustered at neocortical synapses. J Neurosci 22:1328–1337

Kim G, Kandler K (2003) Elimination and strengthening of glycinergic/GABAergic connections during tonotopic map formation. Nat Neurosci 6:282–290

Kotak VC, Fujisawa S, Lee FA, Karthikeyan O, Aoki C, Sanes DH (2005) Hearing loss raises excitability in the auditory cortex. J Neurosci 25:3908–3918

Kotak VC, Breithaupt AD, Sanes DH (2007) Developmental hearing loss eliminates long-term potentiation in the auditory cortex. Proc Natl Acad Sci USA 104:3550–3555

Kundsen EI (2002) Instructed learning in the auditory localization pathway of the barn owl. Nature 417:322–328

Li Y, Fitzpatrick D, White LE (2006) The development of direction selectivity in ferret visual cortex requires early visual experience. Nat Neurosci 9:676–681

Maffei A, Nelson SB, Turrigiano GG (2004) Selective reconfiguration of layer 4 visual cortical circuitry by visual deprivation. Nat Neurosci 7:1353–1359

Maffei A, Nataraj K, Nelson SB, Turrigiano GG (2006) Potentiation of cortical inhibition by visual deprivation. Nature 443:81–84

Mendelson JR, Schreiner CE, Sutter ML, Grasse KL (1993) Functional topography of cat primary auditory cortex: responses to frequency-modulated sweeps. Exp Brain Res 94:65–87

Merzenich MM, Jenkins WM, Johnston P, Schreiner C, Miller SL, Tallal P (1996) Temporal processing deficits of language-learning impaired children ameliorated by training. Science 271:727–781

Misgeld U, Drew G, Yanovsky Y (2007) Presynaptic modulation of GABA release in the basal ganglia. Prog Brain Res 160:245–259

Mize RR (2002) The organization of GABAergic neurons in the mammalian superior colliculus. Prog Brain Res 90:219–248

Mody I (2005) Aspects of the homeostatic plasticity of GABAA receptor-mediated inhibition. J Physiol 562:37–46

Morales B, Choi SY, Kirkwood A (2002) Dark rearing alters the development of GABAergic transmission in visual cortex. J Neurosci 22(18):8084–8090

Murthy VN, Schikorski T, Stevens CF, Zhu Y (2001) Inactivity produces increases in neurotransmitter release and synapse size. Neuron 32:673–682

Nelken I, Versnel H (2000) Responses to linear and logarithmic frequency-modulated sweeps in ferret primary auditory cortex. Eur J NeuroSci 12:549–562

Nugent FS, Penick EC, Kauer JA (2007) Opioids block long-term potentiation of inhibitory synapses. Nature 446:1086–1090

Ortinski PI, Lu C, Takagaki K, Fu Z, Vicini S (2004) Expression of distinct alpha subunits of GABAA receptor regulates inhibitory synaptic strength. J Neurophysiol 92:1718–1727

Ramoa AS, Mower AF, Liao D, Jafri SI (2001) Suppression of cortical NMDA receptor function prevents development of orientation selectivity in the primary visual cortex. J Neurosci 21:4299–4309

Razak KA, Fuzessery ZM (2002) Functional organization of the pallid bat auditory cortex: emphasis on binaural organization. J Neurophysiol 87:72–86

Razak KA, Huang L, Pallas SL (2003) NMDA receptor blockade in the superior colliculus increases receptive field size without altering velocity and size tuning. J Neurophysiol 90:110–119

Razak KA, Pallas SL (2005) Neural mechanisms of stimulus velocity tuning in the superior colliculus. J Neurophysiol 94:3573–3589

Razak KA, Fuzessery ZM (2006) Neural mechanisms underlying selectivity for the rate and direction of frequency-modulated sweeps in the auditory cortex of the pallid bat. J Neurophysiol 96:1303–1319

Razak KA, Pallas SL (2006) Dark rearing reveals the mechanism underlying stimulus size tuning of superior colliculus neurons. Vis Neurosci 23:741–748

Razak KA, Fuzessery ZM (2007) Development of inhibitory mechanisms underlying selectivity for the rate and direction of frequency-modulated sweeps in the auditory cortex. J Neurosci 27:1769–1781

Razak KA, Pallas SL (2007) Inhibitory plasticity facilitates recovery of stimulus velocity tuning in the superior colliculus after chronic NMDA receptor blockade. J Neurosci 27:7275–7283

Razak KA, Fuzessery ZM (2008) Facilitatory mechanisms underlying selectivity for the direction and rate of frequency modulated sweeps in the auditory cortex. J Neurosci 28:9806–9816

Rhoades RW, Chalupa LM (1978a) Receptive field characteristics of superior colliculus neurons and visually guided behavior in dark-reared hamsters. J Comp Neurol 177:17–32

Rhoades RW, Chalupa LM (1978b) Effects of neonatal cortical lesions upon directional selectivity in the superior colliculus of the golden hamster. Brain Res 147:188–193

Roberts EB, Meredith MA, Ramoa AS (1998) Suppression of NMDA receptor function using antisense DNA block ocular dominance plasticity while preserving visual responses. J Neurophysiol 80:1021–1032

Schneider GE (1973) Early lesions of the superior colliculus: Factors affecting the formation of abnormal retinal projections. Brain Behav Evol 8:73–109

Shapley R, Hawken M, Ringach DL (2003) Dynamics of orientation selectivity in the primary visual cortex and the importance of cortical inhibition. Neuron 38:689–699

Shi J, Aamodt SM, Constantine-Paton M (1997) Temporal correlations between functional and molecular changes in NMDA receptors and GABA neurotransmission in the superior colliculus. J Neurosci 17:6264–6276

Smith MA, Bair W, Movshon JA (2006) Dynamics of suppression in macaque primary visual cortex. J Neurosci 26:4826–4834

Sperry RW (1963) Chemoaffinity in the orderly growth of nerve fiber patterns and connections. Proc Natl Acad Sci USA 50:703–710

Stein BE, Dixon JP (1979) Properties of superior colliculus neurons in the golden hamster. J Comp Neurol 183:269–284

Suga N (1965) Analysis of frequency-modulated sounds by auditory neurones of echo-locating bats. J Physiol 179:26–53

Turrigiano GG (1999) Homeostatic plasticity in neuronal networks: the more things change, the more they stay the same. Trends Neurosci 22:221–227

Turrigiano GG, Nelson SB (2004) Homeostatic plasticity in the developing nervous system. Nat Rev Neurosci 5:97–107

Udin SB, Fawcett JW (1988) Formation of topographic maps. Annu Rev Neurosci 11:289–327

Vale C, Schoorlemmer J, Sanes DH (2003) Deafness disrupts chloride transporter function and inhibitory synaptic transmission. J Neurosci 23:7516–7524

Waleszczyk WJ, Wang C, Burke W, Dreher B (1999) Velocity response profiles of collicular neurons: parallel and convergent visual information channels. Neuroscience 93:1063–1076

Walker GA, Ohzawa I, Freeman RD (1999) Asymmetric suppression outside the classical receptive field of the visual cortex. J Neurosci 19:10536–10553

Webb BS, Dhruv NT, Solomon SG, Tailby C, Lennie P (2005) Early and late mechanisms of surround suppression in striate cortex of macaque. J Neurosci 25:11666–11675

Woodin MA, Ganguly K, Poo MM (2003) Coincident pre- and postsynaptic activity modifies GABAergic synapses by postsynaptic changes in Cl⁻ transporter activity. Neuron 39:807–820

Wu GK, Li P, Tao HW, Zhang LI (2006) Nonmonotonic synaptic excitation and imbalanced inhibition underlying cortical intensity tuning. Neuron 52:705–715

Zheng W, Knudsen EI (1999) Functional selection of adaptive auditory space map by $GABA_A$-mediated inhibition. Science 284:962–965

Chapter 6
Postnatal Maturation and Experience-Dependent Plasticity of Inhibitory Circuits in Barrel Cortex

Qian-Quan Sun

Sensory experience drives the refinement of sensory maps in developing adult sensory cortices (Wiesel and Hubel 1974; Stryker 1978; Crair et al. 1998; Feldman and Brecht 2005). Tremendous progress has been made toward understanding the process of maturation of excitatory networks. Cortical *inhibition* has also been shown to play a vital role in the regulation of critical periods for sensory plasticity (Hensch 2005). However, it is unclear whether neocortical inhibitory networks exhibit experience-dependent postnatal maturation. In my laboratory, we employ the so-called "barrel cortex" (Woolsey and Van der 1970) that, represents the individual whiskers on the snout of rodents. The map exhibits plasticity throughout life, in that under- or over-stimulation of a whisker is reflected by contraction or expansion, respectively, of the barrel representing it in the primary somatosensory cortex (Simons and Land 1987). This review focuses on the mechanisms underlying activity-dependent regulation of neocortical inhibitory circuits and the roles of inhibition in somatosensory cortical map plasticity during postnatal development. The focus will be placed on the following questions related to experience-dependent plasticity of neocortical inhibitory networks. (1) How do intrinsic and synaptic properties of inhibitory circuits in barrel cortex change during postnatal maturation? (2) How does sensory stimulation or deprivation affect the maturation of inhibitory circuits? (3) Does the maturation of neocortical inhibitory circuits proceed in an activity-dependent manner or do they develop independently of sensory inputs? (4) What are the molecular and cellular mechanisms that underlie the activity-dependent or -independent maturation of inhibitory networks?

To understand how barrel cortex plasticity happens at a synaptic level, a linkage has to be made between previous sensory experiences in vivo and intracortical synaptic plasticity recorded in vitro (Jiao et al. 2006). Changes in synaptic strength underlying cortical plasticity can be measured from interneurons in the GAD67-GFP mice, after they have been subjected to alterations in whisker experience during their early postnatal development (e.g. Figs. 6.1–6.3, see Simons and Land 1987).

Q.-Q. Sun (✉)
Department of Zoology and Physiology, Graduate Neuroscience Program, University of Wyoming, 1000 E. University Ave, Laramie, WY, 82071, USA
e-mail: neuron@uwyo.edu

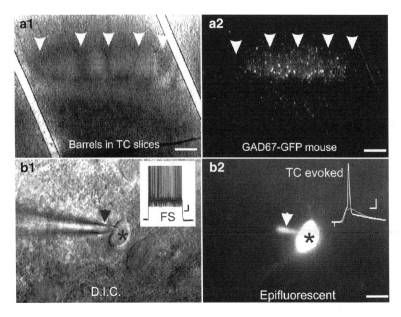

Fig. 6.1 Whole cell recording from interneuron expressing eGFP in barrel cortex of GAD67-GFP mouse. (**a1**) Barrels are visible in a TC slice. White arrowheads: barrel walls. (**a2**) EGFP expressing neurons are abundant in layers 4 barrels. (**b1**) IR-DIC image of a FS neuron. (**b2**) The same neurons visualized under fluorescent microscopy. *: a FS cell

In order to facilitate this approach, we have recently developed a method to obtain dual recordings between excitatory and GAD-GFP-labeled inhibitory neurons in rodent barrel cortex (Jiao et al. 2006; Sun et al. 2006). Individual barrels representing deprived and non-deprived whiskers can be identified in a cortical brain slice preparation, permitting an analysis of the development of excitatory and inhibitory synaptic connections and the underlying synaptic mechanisms that control communication between specific neuronal pairs in vitro (Figs. 6.1 and 6.2). Furthermore, the cortical changes attributable to the selective stimulation of individual whiskers can be investigated.

6.1 Postnatal Maturation and Plasticity of Electrical Properties of Interneurons in the Barrel Cortex

6.1.1 Postnatal Maturation of Electrical Properties in Neocortical Interneurons

Based on electrical properties, interneurons can be separated into three broadly defined groups termed regular spiking (RS), bursting (BS) and fast-spiking (FS) (Connors et al. 1982; McCormick et al. 1985); (Wang et al. 2002; Wang et al. 2004). BS cells

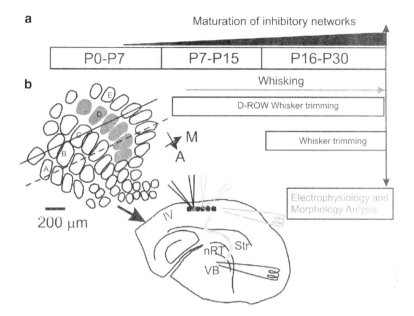

Fig. 6.2 Experimental paradigms for experience dependent plasticities in the barrel cortex. (**a**) Experimental procedure for studying experience-dependent plasticity in barrel cortex. (**b**) Image showing tangential section of layer IV of a flattened barrel cortex. WT was performed for row-D. Sites of intracellular recording and extra cellular stimulation in a thalamocortical brain slice across the barrel field. WT: Whisker trimmig

have also been named low-threshold spiking cells (LTS) (Kawaguchi et al. 1995; Xiang et al. 1998). In addition, there is a subset of FS interneurons that has been characterized as 'irregular spiking, or stuttering' (Ma et al. 2006). Interneurons that are involved in experience-dependent plasticity are likely to be a key component of sensory processing circuits because they modulate temporal and spatial properties of sensory-mediated cortical activities. Agmon and colleagues have shown that diverse groups of interneurons, including both FS and RS inhibitory cells, fired on thalamocortical (TC) stimulation (Porter et al. 2001). They also reported that the characteristic firing patterns of cortical interneurons seen in adults are absent in neonates. In earlier years, David Prince and colleagues documented the electrical properties of immature neocortical neurons of rats. They found that immature cells (including interneurons) have more positive resting potentials, lower spike amplitude and longer spike duration, higher input resistance, and longer membrane time constants (McCormick and Prince 1987; Kriegstein et al. 1987; Luhmann and Prince 1991). Recent studies have focused on developmental changes in specific interneuronal subtypes, as described next.

6.1.1.1 Maturation of FS and RS-Type Firing Phenotypes

Massengill et al. (1997) reported that RS and FS cells are derived from immature multiple-spiking (IMS) neurons. They found that increased expression of the Kv3.1

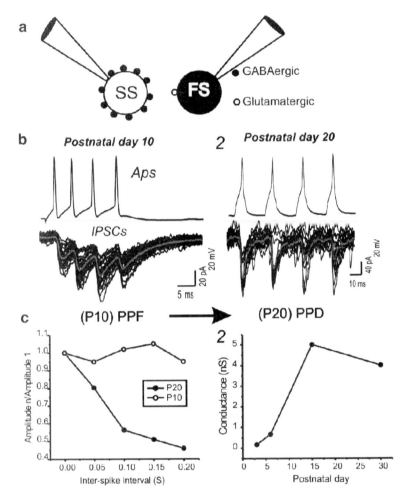

Fig. 6.3 Developmental maturation of GABAergic transmission in barrel cortex layer IV. (**a**) Schematic graph showing paired recordings from SS–FS pair, note that the inhibitory synaptic boutons (from a single basket cell) on a spiny neuron outnumber the glutamatergic boutons from a single spiny neuron on the basket cell (see Sun et al. 2006). (**b**) Paired recording from a SS–FS pair at P10 (**b1**) and P20 (**b2**), uIPSCs elicited by a trin of spikes in the FS neuron show short-term plasticity (**b1**: paired-pulse facilitation [PPF] at P10; **b2**: paired pulse depression [PPD] at P21), APs: action potentials. (**c1**) Relationship between prired pulse ration (PPR) and the inter-spike interval in young (P10, open circles) and older (P21, filled circles) animals. (**c2**) uIPSC conductance recorded from spiny neurons showed developmental maturation

gene contributes to the maturation of electrical phenotypes of FS cells. Other important genes in the maturation of FS firing phenotypes are the Kv3.2 group (Lau et al. 2000). In Kv3.2 knock-out mice, the ability to fire spikes at high frequencies was impaired (Lau et al. 2000). These two studies demonstrated that changes in the expression level of distinct Kv3 channels contribute to postnatal maturation of the electrical properties of FS cells.

6.1.1.2 Maturation of BS or LTS Firing Phenotypes

Interneurons with BS or LTS firing patterns were not recorded in barrel cortex of juvenile animals (Ali et al. 2007), indicating a late maturation of these cells. In a recent study, Connors and colleagues (Long et al. 2005) reported that the synchronous firing among LTS cells was absent at postnatal day 12 (P12) but appeared abruptly shortly after P12. Because developmental transformation of LTS cells into a synchronous, oscillatory network overlaps with the onset of active whisker exploration, Connors and colleagues suggested that there is a potential role for this synchronizing system in development of sensory processing (Long et al. 2005).

6.1.2 Increases in Dendritic Gap Junction (GJ) Coupling During Postnatal Maturation

FS cells form GJ coupled networks (Galarreta and Hestrin 2002; Gibson et al. 2005). Recently, anatomical studies have shown that the GJs are located discretely in the dendrites (Liu and Jones 2003; Fukuda et al. 2006). Connors and colleagues proposed that due to their low pass filtering electrical properties of the GJ, the main functional role for GJs is to effectively propagate small, slow signals, such as afterhyperpolarizations, burst envelopes, or subthreshold oscillations (Mancilla et al. 2007). In barrel cortex, as well as in other neocortical regions, different connexin (Cx) isoforms show distinct maturation patterns. Between postnatal days 0–28, Cx43 and Cx32 expression increases exponentially, whereas Cx26 expression peaks at around P14 (Nadarajah et al. 1996; Nadarajah and Parnavelas 1999; Montoro and Yuste 2004). Overall, the increased expression level of Cx during the second postnatal week of highly enhanced excitatory activity and critical developmental events is consistent with the need for recruitment of interneuronal networks by TC activity and the promotion of spike synchronization in spiny stellate neurons (Sun et al. 2006).

6.1.3 Experience-Dependent Maturation of Electrophysiological Properties of Inhibitory Interneurons

Simons and Land first showed that sensory experiences are crucial for forming normal response properties of single neurons in the adult barrel cortex (Simons and Land 1987).

Simons and colleagues later reported that the firing rates of FS cells are also modified by sensory experiences in vivo. FS inhibitory cells fire less robustly when by whisker-trimming is performed early in life and the re-grown whiskers are stimulated (Lee et al. 2007). They pointed out that it is unknown whether the reduction in sensory-induced layer IV FS interneuron firing is due to changes in intrinsic firing properties, such as increased firing threshold (Barth et al. 2004; Lee et al. 2007), or to synaptic changes, such as reduced intracortical inhibition (Jiao et al. 2006). So far, it remains unclear whether maturation of intrinsic properties of specific interneuron classes undergoes experience-dependent or independent (or both) change. Recently, the term 'intrinsic plasticity' has been used to describe changes in intrinsic firing properties. We have recently been conducting research in this area and our preliminary data indicate that the intrinsic plasticity is cell-type specific, i.e. while FS exhibited intrinsic plasticity in response to whisker trimming, intrinsic properties of RSNP cells does not (Sun QQ, unpublished observations).

6.2 Postnatal Maturation of Intracortical Inhibitory Synaptic Transmission in the Barrel Cortex

In addition to being the major inhibitory neurotransmitter, GABA is thought to play a morphogenetic role in embryonic development. The role of GABA as a trophic factor during neurogenesis at early embryonic stages has already been reviewed in a number of excellent articles (Varju et al. 2001), and thus I only focus on the role of GABA in circuit formation in barrel cortex.

6.2.1 Early Postnatal Development of the GABA System and its Role in Circuit Formation in the Barrel Cortex

6.2.1.1 Synthetic Enzymes for GABA Exhibit Different Expression Patterns

Distinct genes encode two isoforms of the GABA-synthesizing enzyme glutamic acid decarboxylase (GAD: GAD65 and 67). In the barrel cortex, the distribution of GAD65 and GAD67 in the early circuit formation period and the late experience-dependent circuit refinement stage shows a different pattern. Jones and colleagues reported that GAD67 mRNA was highest in layer I at birth and developmentally upregulated in other layers shortly after birth (Golshani et al. 1997). Kiser et al. (1998) showed that between P3 and P6 GAD67-IR coincide with the barrel pattern in layer IV, this pattern are maintained throughout the postnatal period to adulthood. Similar results have also been reported in another study showing that the appearance of GAD67 slightly precedes the onset of barrel formation (Rice and Van der 1977). Thus the enhanced expression of GAD67 from P3-P6 through P9 coincides with the formation of barrels and early critical periods of structural plasticity. In contrast, development of GAD65-IR was delayed relative to GAD67. GAD65-IR, which was scarcely evident before P6, increased markedly in density within cell bodies over the next several weeks. During this prolonged developmental process, GAD65-IR first formed a negative image of the barrels. Later, GAD65-IR was distributed uniformly across layer IV (Kiser et al. 1998a). Based on the above results, Mower and colleagues suggested that the developmental maturation of the barrel cortex involves the following steps: the disappearance of an early GAD67 pattern, mature GAD67 system take over in an inside-outside fashion, and a delayed and prolonged expansion of the GAD65 system (Kiser et al. 1998). Overall, the spatiotemporal differences in postnatal expression of the two GAD isoforms in the barrel cortex indicate different roles of GAD isoforms in early circuit formation and late circuit maturation. It is unclear whether the roles of GAD65 and GAD67 in barrel circuit formation are related to their distinctive contribution to cellular and synaptic GABA levels, respectively.

6.2.1.2 GABA-Mediated Synaptic Transmission in the Early Postnatal Period

I present studies focused on the spatiotemporal relationships between GABA, GAD, GABAR and barrel formation. I will also use this approach to answer questions raised in Sects. 2.2 and 2.3. In barrel cortical neurons from neonatal mice, GABA-mediated IPSPs are recorded as early as postnatal days 0–2. However, the immature postsynaptic potentials (PSPs) are very different from mature IPSPs in reversal potential and latency (Agmon et al. 1996). During postnatal brain development, the reversal potential for $GABA_A$-mediated responses is shifted from −46 mV (postnatal day 0) to −82 mV (>postnatal day 12) (Owens et al. 1999). The upregulation of a K^+-Cl^- coupled co-transporter (KCC_2) is primarily responsible for this shift (Rivera et al. 1999). The patterns of gene expression for the $\alpha1$, $\alpha2$, $\alpha4$, $\alpha5$, $\beta1$, $\beta2$, and $\gamma2$ subunits of mRNAs of $GABA_A$ receptors have also been studied. The $\alpha1$, $\beta2$, and $\gamma2$ subunit mRNAs were highly expressed in the dense cortical plate at birth and increased substantially with age, especially in deep layers (Golshani et al. 1997). Together with the electrophysiology studies, these results suggest that the GABA synthesizing enzymes, specific $GABA_A$ receptors, and GABA-mediated synaptic potentials coexist prior to the formation of visible barrels. Furthermore, the expression of GAD67 and $GABA_A$ receptors showed barrel-like patterns and co-regulated with the barrel formation during development in a similar manner. Therefore, GABA and its $GABA_A$-mediated depolarizing signals may play a role in the early formation of barrel circuits, however, a causal relationship between GABA and barrel formation is yet to be established.

6.2.2 Late Postnatal and Experience-Dependent Maturation of Inhibitory Circuits in the Barrel Cortex

6.2.2.1 Presynaptic Maturation

Experience-dependent synaptic plasticity requires precise timing between pre and postsynaptic excitatory cortical neurons (Feldman and Brecht 2005). Intracortical inhibition promotes the temporal precision of information relay by shunting recurrent cortical excitation. This idea is supported by recordings in vivo in the somatosensory and other sensory cortices (Moore and Nelson 1998; Kelly et al. 1999; Bruno and Simons 2002). To serve a role in experience-dependent plasticity of neural circuits, the weight of inhibitory synapses must be regulated during postnatal period. Indeed this is the case, for example, enhancing whisker activity increases the number of GABAergic synapses formed on dendritic spines (Knott et al. 2002). On the other hand, regulation of NMDA receptor subtype composition has no effect on the critical period for barrel formation (Lu et al. 2001a). In the barrel cortex, GAD65 expression increases late in the critical period (Kiser et al. 1998b). Together, these evidences support GABA's role in the refinement of barrel structure. Additional experiments that thoroughly examine the roles of GABA in barrel plasticity are necessary for a more complete understanding of the roles of inhibition in cortical development.

6.2.2.2 Postsynaptic maturation

In an in situ hybridization study, $GABA_A$ receptor subunits ($\alpha 1$, $\beta 2$, $\beta 1$ and $\gamma 2$) increased substantially with age in the barrel circuits (Golshani et al. 1997). A patch clamp study by Agmon and colleagues also noted an increase in conductance of evoked $GABA_A$ PSPs in the first postnatal week (Agmon et al. 1996). We (Jiao et al. 2006) have recently found similar results, in addition, we showed that the presynaptic properties (e.g. quantal content and paired-pulse properties) of IPSCs of immature (P7) neurons are different from mature cells (P30, cf. Fig. 6.3).

6.2.2.3 Experience-Dependent Postnatal Maturation

Simons and Land first reported that functional plasticity is a fundamental aspect of cortical development in barrel cortex (Simons and Land 1987). In a subsequent study, Woolsey, McCasland and colleagues reported important role of afferent sensory activities in the structural maturation of cortical circuits (McCasland et al. 1992). They found that local cortical axons (excitatory and inhibitory) do not mature after early deafferentation. Recent studies focused specifically on GABAergic transmission. Sun et al. (Sun et al. 2006) studied intracortical inhibitory transmission onto spiny stellate cells in rat TC slices. We reported that unitary conductances of IPSCs produced by a single FS cell are about 10 times larger than unitary conductances of excitatory neurons and are 10 nS in P20-P35 animals (Sun et al. 2006). Interestingly, in sensory-deprived mature barrel cortex, the properties of evoked and miniature IPSCs in mature brain are similar to those recorded in immature brain (e.g. Fig. 6.3) (Jiao et al. 2006; Sun et al. 2006). In summary, these results suggest that GABAergic synaptic transmission undergoes rapid developmental maturation and that this process is fine tuned by sensory experience.

6.2.3 Interneurons involved in sensory feed-forward inhibition in the barrel cortex and the consequences of their functional maturation to network processing

Strong and reliable unitary feed-forward inhibition onto excitatory neurons in layer IV serves to effectively "shunt" recurrent excitation and preserve discrete signaling in cortical networks (Castro-Alamancos 2000; Wilent and Contreras 2005; Cruikshank et al. 2007). Swadlow and colleagues were the first to propose that FS interneurons are major candidates for providing feed-forward inhibition (Swadlow 2002, 2003). Using paired recording techniques in thalamocortical (TC) slices, we (Sun et al. 2006) tested Swadlow's proposition by examining interactions between synaptically connected excitatory and inhibitory neurons in layer IV of barrel cortex. We demonstrated that small clusters of FS cells can be reliably and precisely activated by TC inputs and provide feed-forward inhibition onto excitatory neurons (Sun et al. 2006). Connors and colleagues elucidated the synaptic mechanisms underlying selective activation of layer IV FS interneurons (Cruikshank et al. 2007). They found that synaptic mechanisms are responsible for the greater responsiveness in interneurons vs. excitatory cells. As a result, response properties of excitatory neurons correlate well with sensory inputs and thus allow spike-timing dependent plasticity. In the neonate, GABA is depolarizing and believed to have a different role than in adults. How does the transformation of the functional role of inhibitory GABAergic transmission occur in barrel cortex? Issac and colleagues (Daw et al. 2007) showed that the $GABA_A$ receptor conductance is depolarizing in neonates (postnatal days 3–5), but GABAergic transmission at this age is not elicited by TC input and has no detectable circuit function. However, recruitment of GABA synapses occurs at the end of first postnatal period as a result of coordinated increases in TC drive to FS cells. Thus, GABAergic circuits are not engaged by TC input in the neonate, but are abruptly involved in the feed-forward inhibitory circuit at the end of the first postnatal week (Daw et al. 2007). Surprisingly, this transformation occurs apparently coincidentally within the time window of the disappearance of silent synapses (i.e. synapses only exhibiting NMDA receptor-mediated responses) and the critical period for TC dependent long-term glutamatergic synaptic plasticity (Feldman et al. 1999). A logic step toward future investigation is to understand how such an abrupt maturation occur at specific GABAergic cells.

6.3 Does the Maturation of Neocortical Inhibitory Networks Proceed in an Activity-Dependent Manner or Independently of Sensory Inputs (or Both)?

As in visual cortex, early postnatal sensory experiences are crucial for forming mature functional cortical circuits in the barrel cortex (McCasland et al. 1992). In the visual cortex, deletion of synaptic GAD (i.e. GAD65) can abolish critical periods (Hensch and Stryker 2004; reviewed by Hensch 2005). How does inhibition contribute to the initiation and closure of the neocortical critical periods? A very compelling hypothesis about the role of inhibition in the initiation and closure of critical periods is that it can modulate Hebbian-type plasticity (Hebb 1955) by enhancing correlative neuronal firing among adjacent cells and anti-correlative firing in distal cells (Hensch and Stryker 2004). To serve this role, i.e. modulating the spike-timing and lateral spread of excitation, the strength of inhibitory synapses needs to be developmentally regulated as well. Prior to the closure of neocortical critical periods, TC and intracortical glutamatergic synapses undergo drastic morphological, molecular and functional changes (Feldman et al. 1998). Disturbances in the balance of excitation and inhibition in the neocortex induce cortical epileptic seizure (Prince 1999). Therefore, a key requirement for the maturation of sensory cortices, based on a Hebbian-rule, is that excitation and inhibition must be delicately balanced to achieve appropriate functioning at the level of local cortical microcircuit.

6.3.1 Experience-Dependent Plasticity of GABAergic Circuits in the Barrel Cortex

In the barrel cortex, there is considerable evidence suggesting that the amount of inhibitory neurotransmitter (GABA), GABA receptors, and the number of GABAergic synapses are correlated with levels of neuronal activity (Micheva and Beaulieu 1997; Jiao et al. 2006; Knott et al. 2006). Here, I review studies focused on the effects of whisker trimming or stimulation on inhibitory circuits of the barrel cortex.

6.3.1.1 Sensory Deprivation (Whisker-Trimming)

In vivo electrophysiological studies: In an earlier study, it was shown that whisker removal produces immediate disinhibition in the neighboring whisker barrels (Kelly et al. 1999b). In more recent studies, Simons and colleagues examined how this process is regulated in cortices representing the trimmed whiskers. They reported that excitatory neurons in deprived barrels displayed higher spontaneous firing rates, more robust responses to whisker stimulation, and weaker inhibitory interactions between neurons representing neighboring whiskers (Simons and Land 1987;

Silberberg et al. 2004). In contrast, recordings from FS neurons indicate that these cells fire less robustly under the same conditions (Shoykhet et al. 2005; Lee et al. 2007). More intriguingly, the deprivation effects persist even after months of whisker re-growth, (Shoykhet et al. 2005). These result suggests that whisker-dependent structural alterations may have occurred in cortical circuits during postnatal developmental period. Similar effects (i.e. disinhibition) are seen in deafferented developing visual and auditory centers (Pallas et al. 2006; Razak and Pallas 2006). Simons and colleagues proposed that the contrasting effects in excitatory and inhibitory neurons may reflect altered patterns of TC input to excitatory versus inhibitory cells or changes in the strength of intracortical connections.

In vitro electrophysiological and neuroanatomical studies: In an earlier study on barrel cortex in rats, Micheva and Beaulieu showed that unilateral whisker trimming induces highly selective changes in cortical GABA circuitry of both hemispheres (Micheva and Beaulieu 1995). As indicated earlier in this chapter (Figs. 6.1–6.3), brain slice preparations allow a linkage to be made between synaptic properties recorded *in vitro* and previous sensory experiences *in vivo*. Using this approach, we (Jiao et al. 2006) showed that row D whisker trimming begun at P7, but not after P15, induced a reduction in the number of inhibitory perisomatic varicosities, and reduced synaptic GAD65/67 immunoreactivity in spiny neurons of the deprived barrels (Fig. 6.4). Patch-clamp recording from

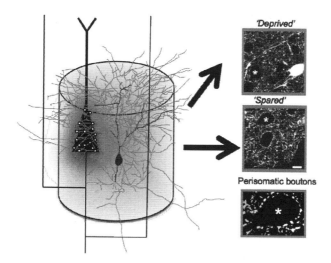

Fig. 6.4 Perisomatic GABAergic innervation of excitatory neurons by basket cells is modified by sensory experience in vivo. A reconstructed basket cell (FS type) recorded from barrel cortex layer IV (Q.Q., Sun). The location of the FS cells in the barrel (*cylinder*) is shown. Note that the axons (*gray*) of the basket cell are predominantly confined within the barrel. Perisomatic GABergic synaptic contacts (*white circles* and digitally enhanced micrograph in the lower right corner) were formed in the some area of the principal neuron (a star pyramidal neuron, *black triangly*). Whisker trimming induced reduction of the number of perisomatic boutons (two micrographs on the right)

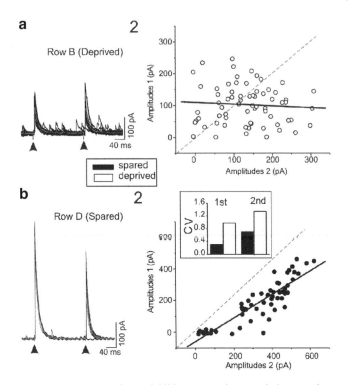

Fig. 6.5 Effects of sensory deprivation on inhibitory synaptic transmission on spiny neurons. (**a1**) IPSCs were recorded in a spiny newron located in the 'deprived' row. the IPSCs were evoked by an adjacent extracellular stimulating electrode. (**a2**) The amplitudes of the second evoked IPSCs were plotted against the amplitudes of the first IPSCs. Solid line: linear regression fit. (**b1**) IPSCs were evoked in a spiny neuron located in a 'spared' row. (**b2**) Scatter plot of the amplitudes of IPSC1 vs. IPSC2. Solid line: linear regression fit. CV: Coefficient of Vairance for the evoked IPSCs (first evoked IPSCs, second evoked IPSCs). This figure was modified from Jiao et al. 2006

spiny cells showed a 1.5-fold reduction of intracortical evoked IPSCs (eIPSCs) in deprived versus spared cortices (Fig. 6.5). The reduction in eIPSCs occurred via changes in presynaptic properties (i.e. quantal content, paired pulse ratio and synaptic numbers) and unitary IPSC amplitudes (Fig. 6.5). Miniature IPSCs showed subtle but significant differences in the quantal amplitudes between the two experimental conditions. In addition, properties of the IPSCs in deprived barrels of adults resembled those of IPSCs recorded in immature brains (P7). We concluded that the perisomatic inhibition mediated by PV-positive basket cells is pruned by sensory deprivation (Jiao et al. 2006). In addition, the dendritic GABAergic synapses were reduced in number with sensory deprivation (Micheva and Beaulieu 1995). Together, these results strongly suggest that the properties of local intracortical inhibitory networks are modified by sensory experience.

6.3.1.2 Whisker Stimulation

Combining high-resolution 2-deoxyglucose (2DG) and immunohistochemical staining for GABA specific antibodies, McCasland and Hibbard (1997), McCasland et al. (1997) reported that putative inhibitory neurons in barrel cortex of behaving animals are much more heavily labeled than presumed excitatory cells. This metabolic activation is dependent specifically on sensory inputs from the whiskers, because acute trimming of most whiskers greatly reduces 2DG labeling in both cell classes in columns corresponding to trimmed whiskers (McCasland and Hibbard 1997; McCasland et al. 1997). In addition, the same group has reported that PV cells were metabolically more active than other interneurons (Maier and McCasland 1997). In a histological study, Knott et al. (2006) reported that chronic stimulation of a mystacial whisker follicle induces structural and functional changes in layer IV of the corresponding barrel. The changes include insertion of new inhibitory synapses onto spines in an excitatory cell and a depression of neuronal firing rate to the stimulated whisker. In another anatomical study, active whisking was found to accelerate the appearance of mature inhibition (Kiser et al. 1998a). Welker and colleagues (Quairiaux et al. 2007) analyzed how sensory responses of single units are affected in different layers of the stimulated and adjacent barrel columns. They reported that an increased inhibition within the stimulated barrel, a reduction of flow of excitation toward superficial layers and reduction of subsequent spread of excitation toward adjacent columns (Quairiaux et al. 2007). The opposing effects of whisker stimulation (Knott et al. 2006; Quairiaux et al. 2007) compared to whisker trimming (Shoykhet et al. 2005; Jiao et al. 2006) on strength of intracortical inhibitory networks suggest that the strength of intracortical inhibition is fine tuned to balance the amount of intracortical excitation during the critical periods of postnatal development. Disturbances in the activity pattern shift the balance of inhibition and excitation to facilitate the functional and structural lateral intracortical re-organization.

6.3.2 *Activity-Independent Maturation and Plasticity of GABAergic Circuits*

Activity-independent mechanisms regulate mainly postsynaptic aspects of network maturation. In addition to clearly defined activity-dependent processes that underlie GABAergic maturation described above, activity-independent plasticity has been reported in sensory cortices by determining what aspects of maturation occur despite deafferentation. In the barrel cortex, the density of $GABA_A$ receptors is reduced in layer IV following complete loss of peripheral afferent input. However, less severe tactile deprivation had little or no effect on $GABA_A$ receptor distribution (Land et al. 1995). In a similar study, Fuchs and Salazar (1998) reported that intact whisker input is not required for the developmental increase in GABA(A) receptors.

These results are similar to a result obtained in the visual cortex, where a lack of extrinsic input to the visual cortex does not affect the overall developmental regulation of synaptic functioning of $GABA_A$ receptors (Heinen et al. 2004). In a few studies, both activity-dependent and -independent mechanisms were shown to contribute to GABAergic maturation. Itami et al. (2007) reported that the characteristic electrophysiological properties of FS cells were underdeveloped or did not appear at all in BDNF(−/−) mice. Similar results have been reported in the visual cortex, where over-expression of BDNF promotes the maturation of GABA transmission in the absence of activity (via dark rearing) in the visual cortex (Gianfranceschi et al. 2003) and other cortical regions (see reviews by Lu et al. 2005; Woo and Lu 2006). Thus, neurotrophic factors such as BDNF appear to regulate the maturation of the GABAergic system in an activity-independent manner. However, the transcription of BDNF gene is controlled by four promoters, which drive the expression of four transcripts coding for the same protein. Promoter-IV mediates activity-dependent BDNF transcription. It remains unclear how these different transcriptional components contribute to the total effects of BDNF during postnatal cortical maturation. Future studies, using refined molecular approaches to selectively silent a specific BDNF transcriptional pathway (e.g. Promoter-IV) will help to determine whether BDNF acts in parallel with or mediates the activity-dependent regulation of cortical circuits in vivo (Lu B and Sun QQ, unpublished observations. Review by Lu 2003). In summary, a thorough understanding of postnatal maturation process requires not only knowledge of how these different components (pre- vs. postsynaptic) of the GABAergic system change during maturation, but also how they interact with a variety of environmental factors and neurotrophic factors.

6.4 Molecular Mechanisms Underlying Experience-Dependent Plasticity of Inhibitory Circuits in the Barrel Cortex

6.4.1 The Roles of Metabotropic and Ionotropic Glutamate Receptors

6.4.1.1 N-Methyl-D-Aspartate Receptors (NMDARs)

Like experience-dependent plasticity in excitatory networks, NMDARs appear to play an important role in the plasticity of GABAergic synapses. However, direct evidence linking sensory-specific activation of NMDARs with maturation of specific GABAergic circuits is lacking. The cellular mechanisms by which NMDARs regulate GABAergic synapses also appear to differ from those observed in excitatory synapses, in that their actions may take place in presynaptic terminals (Fiszman et al. 2005). In the developing Xenopus retinotectal system, repetitive stimulation of the optic nerve induces LTP of excitatory inputs, but LTD of inhibitory inputs (Lien et al. 2006). The LTD is due to a reduction in presynaptic GABA release and requires activation of presynaptic NMDARs and simultaneous high-level GABAergic activity.

Thus, the presynaptic NMDAR may function as a coincidence detector for adjacent glutamatergic and GABAergic activities, leading to coordinated synaptic modification by sensory experience. In the barrel cortex, in a few studies in which NMDA receptor subunits were knocked out NR1, (Iwasato et al. 1997; Iwasato et al. 2000); NR2A, (Lu et al. 2001), it was shown that intact cortical NMDARs are essential for the aggregation of layer IV cells into barrels and for the development of the full complement of TC patterning, however, there was no effect of a loss of NR2A on the critical periods in barrel cortex (Lu et al. 2001). It remains to be determined whether there is any contribution of specific NMDA receptors to the experience-dependent plasticity of inhibitory cortical networks. To achieve this goal, the next step involves characterization of NMDARs in developing interneurons, for example, developmental switch of NMDAR subunits (e.g. NR2A, NR2B) has been documented in excitatory neurons, whether a similar switch exists in specific cortical interneurons remains to be determined.

6.4.1.2 Metabotropic Glutamate Receptors (mGluRs)

In a recent study (Liu et al. 1998), $mGluR_{1a}$, $mGluR_{5}$, and $mGluR_{2/3}$ were found to be concentrated in layer IV of barrel cortex, particularly in the barrel hollows. This pattern peaks between P4 and P9, a time when intense $NMDAR_1$-IR was also present (cf. (Rema and Ebner 1996)). This finding supports the involvement of mGluRs in the developmental plasticity of TC synapses during the establishment of the somatotopic whisker maps in SI. In addition, an interaction between mGluRs and NMDARs has been demonstrated (Liu et al. 1998). A key component of this interaction may result from synergistic changes in intracellular calcium signaling. For example, mGluRs, via the phospholipase C-b1 (PLC-b1) signaling pathway, regulate intracellular calcium signaling. Indeed, in both PLC-b1 and $mGluR_5$ knockout mice, barrel formation was disrupted (Spires et al. 2005). Expression of several mGluR isoforms has been reported in GABA releasing interneurons in neocortex (Baude et al. 1993; Lujan et al. 1997; Dalezios et al. 2002). However, the exact role of specific mGluRs in regulation of sensory-dependent plasticity of inhibitory circuits remains to be determined. In a recent study (Sun et al. 2009), our group investigated cell specific expression and modulation by mGluRs. We found that whereas activation of group I, II and III mGluRs inhibited glutmatergic transmission in RSNP interneurons, group I mGluR activation depolarizes FS cells only. Thus, there are cell-type and circuit specific roles for mGluR in modulation and plasticity of inhibitory circuits.

6.4.2 Transcriptional Factors and Maturation of Inhibitory Circuits

Activity-dependent signaling pathways induce neuronal gene transcription by modulating transcriptional activators and repressors that are important for neuronal survival and differentiation, synaptogenesis, and plasticity (West et al. 2002). It is now generally agreed upon that activity – transcription coupling is an important step leading to permanent plastic changes in neuronal structure and function. Recent work has shown that sensory information processing is accompanied by the induction of several transcription factors in the barrel cortex. Using in situ hybridization, several groups investigated the effects of whisker stimulation in freely moving rats on the expression of immediate-early genes in the barrel cortex. These studies have consistently reported enhanced *zif 268* and c-*fos* expression that was largely restricted to radial columns across the barrels representing the stimulated whiskers, especially in layer IV. They reported that the majority of activated cells are excitatory, however, GABAergic interneurons were also activated (Filipkowski et al. 2000, 2001; Staiger et al. 2002). A number of studies also indirectly addressed the issue of activation of transcriptional factors and maturation of inhibitory circuits. For example, BDNF has been reported to be critical for the development of cortical inhibitory neurons. In a recent gene expression profiling study using oligonucleotide microarrays performed in cortical tissue from mice with inducible deletions of BDNF, Glorioso et al. (Glorioso et al. 2006) studied the role of BDNF in the expression of transcripts whose protein products are involved in GABA transmission. In this study, loss of BDNF in both embryonic and adult stages gave rise to many shared transcriptome changes. BDNF appeared to be required to maintain gene expression in the SST-NPY-TAC1 subclass of GABA neurons. They have observed BDNF-dependent alterations in genes encoding early-immediate genes (ARC, EGR1, EGR2, FOS, DUSP1, DUSP6) and critical cellular signaling systems (CDKN1c, CCND2, CAMK1g, RGS4) (Glorioso et al. 2006). However, it is unclear which component of these BDNF-dependent gene expression changes is involved in the activity-dependent transcriptome changes underlying experience-dependent plasticity of barrel inhibitory circuits.

6.4.3 The Roles of GABA and GAD

$GABA_A$ agonist infusion in visual cortex *in vivo* restores critical period formation in GAD65 knockout mice (Fagiolini et al. 2003; Hensch et al. 1998). Suppressing GABA reuptake or applying $GABA_A$ agonist in cultured cortical neurons can rescue cell autonomous deficits in axon branching and synapse formation (Chattopadhyaya et al. 2007). These studies indicate that GAD-mediated GABA synthesis regulates the formation of inhibitory synapses in pyramidal neurons. In the barrel cortex, GAD expression decreases during deprivation (Akhtar and Land 1991) and increases following sensory stimulation (Welker et al. 1989), consistent with the idea that GABA and its enzyme GAD is actively involved in activity-dependent maturation of inhibitory circuits. Because the whisker trimming-induced plasticity occurs in the second to fourth postnatal weeks, the $GABA_A$-mediated currents during this period are already hyperpolarizing. How GABA contributes to the activation of transcriptional regulation remains to be determined.

6.5 Concluding Remarks

Studies carried out in the barrel cortex complement the progress made in the auditory and visual systems. They have provided additional new insights into the role of inhibition in remodeling neural circuits in neocortical layer IV and II/III and the role of sensory experience in the maturation of GABAergic networks. The following paragraph is a biased summary of major progress made in this regard and the remaining issues that need to be addressed in future studies. (1) Inhibition plays a critical role in the experience-dependent refinement of cortical circuits. However, the mechanisms underlying the maturation of inhibitory networks and plasticity at GABAergic synapses have remained elusive. Current evidence indicates that, like the maturation of excitatory circuits, synaptic reorganization by elimination and strengthening also occurs at GABAergic synapses. However, it is largely unknown how this process is regulated (other chapters address this issue). For example, in the barrel cortex, it is unknown whether there is a critical period for experience-dependent regulation of inhibitory synapses. Information on the role of experience in the maturation of inhibitory networks in the supragranular layers (II/III) and layer V is not available. (2) While inhibition is known to be critical for visual cortex reorganization, much less is known about how sensory experience modifies the structure and function of inhibitory networks themselves. Recent studies (Micheva and Beaulieu 1997; Jiao et al. 2006; Knott et al. 2006) using the barrel sensory system and examining the effects of sensory deprivation or enhancement suggest that the strength of intracortical inhibition is fine tuned to balance the amount of intracortical excitation during the critical periods of postnatal development. Disturbances in the activity pattern shift the balance of inhibition and excitation to facilitate the functional and structural intracortical re-organization. However, information about mechanisms underlying the experience-dependent plasticity of inhibitory circuits is still sketchy at this point. Although several key players has been identified (e.g. BDNF, NMDA receptors, mGluRs, and GABA), how these different players act in concert to modulate the experience-dependent plasticity that occurrs in vivo is unknown. (3) At the network level, the consequences of reduced intracortical inhibitory synaptic transmission upon sensory deprivation to the columnar propagation of sensory-mediated activities have not been completely understood. Implementation of computational simulations, which can incorporate results obtained in electrophysiological studies, can lead to a better understanding about network mechanisms. (4) Finally, a more direct way to assess the roles of GABA in promoting the maturation of barrel circuits would be to manipulate the level of GABA within circuits. This could be done using a combination of pharmacological means and genetic manipulation of GAD. Such experiments would yield important insights into the role of GABA in the formation and plasticity of barrel circuits.

Acknowledgments I thank Chunzhao Zhang, Yuanyuan Jiao, Leah Selby and Andrew Young for their help and excellent assistance in all studies described in this chapter. I thank Dr. Yuchio Yanagawa at the Department of Genetic and Behavioral Neuroscience, Gunma University Graduate School of Medicine for the generous gift of GAD67-GFP mouse. My work is supported by NIH.

Most of work dealing with the properties of excitatory neurons and excitatory synapses in the barrel cortex could not be cited here due to the focus of this book on the GABAergic system. I apologize to my colleagues for such necessary omissions.

References

Agmon A, Hollrigel G, O'Dowd DK (1996) Functional GABAergic synaptic connection in neonatal mouse barrel cortex. J Neurosci 16:4684–4695

Ali AB, Bannister AP, Thomson AM (2007) Robust correlations between action potential duration and the properties of synaptic connections in layer 4 interneurones in neocortical slices from juvenile rats and adult rat and cat. J Physiol 580:149–169

Barth AL, Gerkin RC, Dean KL (2004) Alteration of neuronal firing properties after in vivo experience in a FosGFP transgenic mouse. J Neurosci 24:6466–6475

Baude A, Nusser Z, Roberts JD, Mulvihill E, McIlhinney RA, Somogyi P (1993) The metabotropic glutamate receptor (mGluR1 alpha) is concentrated at perisynaptic membrane of neuronal subpopulations as detected by immunogold reaction. Neuron 11:771–787

Bruno RM, Simons DJ (2002) Feedforward mechanisms of excitatory and inhibitory cortical receptive fields. J Neurosci 22:10966–10975

Castro-Alamancos MA (2000) Origin of synchronized oscillations induced by neocortical disinhibition in vivo. J Neurosci 20:9195–9206

Connors BW, Gutnick MJ, Prince DA (1982) Electrophysiological properties of neocortical neurons in vitro. J Neurophysiol 48:1302–1320

Crair MC, Gillespie DC, Stryker MP (1998) The role of visual experience in the development of columns in cat visual cortex. Science 279:566–570

Cruikshank SJ, Lewis TJ, Connors BW (2007) Synaptic basis for intense thalamocortical activation of feedforward inhibitory cells in neocortex. Nat Neurosci 10:462–468

Dalezios Y, Lujan R, Shigemoto R, Roberts JD, Somogyi P (2002) Enrichment of mGluR7a in the presynaptic active zones of GABAergic and non-GABAergic terminals on interneurons in the rat somatosensory cortex. Cereb Cortex 12:961–974

Daw MI, Ashby MC, Isaac JT (2007) Coordinated developmental recruitment of latent fast spiking interneurons in layer IV barrel cortex. Nat Neurosci 10:453–461

Feldman DE, Brecht M (2005) Map plasticity in somatosensory cortex. Science 310:810–815

Feldman DE, Nicoll RA, Malenka RC (1999) Synaptic plasticity at thalamocortical synapses in developing rat somatosensory cortex: LTP, LTD, and silent synapses. J Neurobiol 41:92–101

Feldman DE, Nicoll RA, Malenka RC, Isaac JT (1998) Long-term depression at thalamocortical synapses in developing rat somatosensory cortex. Neuron 21:347–357

Filipkowski RK, Rydz M, Berdel B, Morys J, Kaczmarek L (2000) Tactile experience induces c-fos expression in rat barrel cortex. Learn Mem 7:116–122

Filipkowski RK, Rydz M, Kaczmarek L (2001) Expression of c-Fos, Fos B, Jun B, and Zif268 transcription factor proteins in rat barrel cortex following apomorphine-evoked whisking behavior. Neuroscience 106:679–688

Fiszman ML, Barberis A, Lu C, Fu Z, Erdelyi F, Szabo G, Vicini S (2005) NMDA receptors increase the size of GABAergic terminals and enhance GABA release. J Neurosci 25:2024–2031

Fuchs JL, Salazar E (1998) Effects of whisker trimming on GABA(A) receptor binding in the barrel cortex of developing and adult rats. J Comp Neurol 395:209–216

Fukuda T, Kosaka T, Singer W, Galuske RA (2006) Gap junctions among dendrites of cortical GABAergic neurons establish a dense and widespread intercolumnar network. J Neurosci 26:3434–3443

Galarreta M, Hestrin S (2002) Electrical and chemical synapses among parvalbumin fast-spiking GABAergic interneurons in adult mouse neocortex. Proc Natl Acad Sci U S A 99:12438–12443

Gianfranceschi L, Siciliano R, Walls J, Morales B, Kirkwood A, Huang ZJ, Tonegawa S, Maffei L (2003) Visual cortex is rescued from the effects of dark rearing by overexpression of BDNF. Proc Natl Acad Sci U S A 100:12486–12491

Gibson JR, Beierlein M, Connors BW (2005) Functional properties of electrical synapses between inhibitory interneurons of neocortical layer 4. J Neurophysiol 93:467–480

Glorioso C, Sabatini M, Unger T, Hashimoto T, Monteggia LM, Lewis DA, Mirnics K (2006) Specificity and timing of neocortical transcriptome changes in response to BDNF gene ablation during embryogenesis or adulthood. Mol Psychiatry 11:633–648

Golshani P, Truong H, Jones EG (1997) Developmental expression of GABA(A) receptor subunit and GAD genes in mouse somatosensory barrel cortex. J Comp Neurol 383:199–219

Heinen K, Bosman LW, Spijker S, van Pelt J, Smit AB, Voorn P, Baker RE, Brussaard AB (2004) GABAA receptor maturation in relation to eye opening in the rat visual cortex. Neuroscience 124:161–171

Hensch TK (2005) Critical period plasticity in local cortical circuits. Nat Rev Neurosci 6:877–888

Hensch TK, Stryker MP (2004) Columnar architecture sculpted by GABA circuits in developing cat visual cortex. Science 303:1678–1681

Itami C, Kimura F, Nakamura S (2007) Brain-derived neurotrophic factor regulates the maturation of layer 4 fast-spiking cells after the second postnatal week in the developing barrel cortex. J Neurosci 27:2241–2252

Iwasato T, Datwani A, Wolf AM, Nishiyama H, Taguchi Y, Tonegawa S, Knopfel T, Erzurumlu RS, Itohara S (2000) Cortex-restricted disruption of NMDAR1 impairs neuronal patterns in the barrel cortex. Nature 406:726–731

Iwasato T, Erzurumlu RS, Huerta PT, Chen DF, Sasaoka T, Ulupinar E, Tonegawa S (1997) NMDA receptor-dependent refinement of somatotopic maps. Neuron 19:1201–1210

Jiao Y, Zhang C, Yanagawa Y, Sun QQ (2006) Major effects of sensory experiences on the neocortical inhibitory circuits. J Neurosci 26:8691–8701

Kawaguchi Y, Wilson CJ, Augood SJ, Emson PC (1995) Striatal interneurones: chemical, physiological and morphological characterization. Trends Neurosci 18:527–535

Kelly MK, Carvell GE, Kodger JM, Simons DJ (1999) Sensory loss by selected whisker removal produces immediate disinhibition in the somatosensory cortex of behaving rats. J Neurosci 19:9117–9125

Kiser PJ, Cooper NG, Mower GD (1998) Expression of two forms of glutamic acid decarboxylase (GAD67 and GAD65) during postnatal development of rat somatosensory barrel cortex. J Comp Neurol 402:62–74

Knott GW, Holtmaat A, Wilbrecht L, Welker E, Svoboda K (2006) Spine growth precedes synapse formation in the adult neocortex in vivo. Nat Neurosci 9:1117–1124

Knott GW, Quairiaux C, Genoud C, Welker E (2002) Formation of dendritic spines with GABAergic synapses induced by whisker stimulation in adult mice. Neuron 34:265–273

Kriegstein AR, Suppes T, Prince DA (1987) Cellular and synaptic physiology and epileptogenesis of developing rat neocortical neurons in vitro. Brain Res 431:161–171

Land PW, de Blas AL, Reddy N (1995) Immunocytochemical localization of GABAA receptors in rat somatosensory cortex and effects of tactile deprivation. Somatosens Mot Res 12:127–141

Lau D, Vega-Saenz de Miera EC, Contreras D, Ozaita A, Harvey M, Chow A, Noebels JL, Paylor R, Morgan JI, Leonard CS, Rudy B (2000) Impaired fast-spiking, suppressed cortical inhibition, and increased susceptibility to seizures in mice lacking Kv3.2 K+ channel proteins. J Neurosci 20:9071–9085

Lee SH, Land PW, Simons DJ (2007) Layer- and cell-type-specific effects of neonatal whisker-trimming in adult rat barrel cortex. J Neurophysiol 97:4380–4385

Lien CC, Mu Y, Vargas-Caballero M, Poo MM (2006) Visual stimuli-induced LTD of GABAergic synapses mediated by presynaptic NMDA receptors. Nat Neurosci 9:372–380

Liu XB, Jones EG (2003) Fine structural localization of connexin-36 immunoreactivity in mouse cerebral cortex and thalamus. J Comp Neurol 466:457–467

Liu XB, Munoz A, Jones EG (1998) Changes in subcellular localization of metabotropic glutamate receptor subtypes during postnatal development of mouse thalamus. J Comp Neurol 395:450–465

Long MA, Cruikshank SJ, Jutras MJ, Connors BW (2005) Abrupt maturation of a spike-synchronizing mechanism in neocortex. J Neurosci 25:7309–7316

Lu B, Pang PT, Woo NH (2005) The yin and yang of neurotrophin action. Nat Rev Neurosci 6:603–614

Lu HC, Gonzalez E, Crair MC (2001) Barrel cortex critical period plasticity is independent of changes in NMDA receptor subunit composition. Neuron 32:619–634

Luhmann HJ, Prince DA (1991) Postnatal maturation of the GABAergic system in rat neocortex. J Neurophysiol 65:247–263

Lujan R, Roberts JD, Shigemoto R, Ohishi H, Somogyi P (1997) Differential plasma membrane distribution of metabotropic glutamate receptors mGluR1 alpha, mGluR2 and mGluR5, relative to neurotransmitter release sites. J Chem Neuroanat 13:219–241

Ma Y, Hu H, Berrebi AS, Mathers PH, Agmon A (2006) Distinct subtypes of somatostatin-containing neocortical interneurons revealed in transgenic mice. J Neurosci 26:5069–5082

Maier DL, McCasland JS (1997) Calcium-binding protein phenotype defines metabolically distinct groups of neurons in barrel cortex of behaving hamsters. Exp Neurol 145:71–80

Mancilla JG, Lewis TJ, Pinto DJ, Rinzel J, Connors BW (2007) Synchronization of electrically coupled pairs of inhibitory interneurons in neocortex. J Neurosci 27:2058–2073

Massengill JL, Smith MA, Son DI, O'Dowd DK (1997) Differential expression of K4-AP currents and Kv3.1 potassium channel transcripts in cortical neurons that develop distinct firing phenotypes. J Neurosci 17:3136–3147

McCasland JS, Hibbard LS (1997) GABAergic neurons in barrel cortex show strong, whisker-dependent metabolic activation during normal behavior. J Neurosci 17:5509–5527

McCasland JS, Hibbard LS, Rhoades RW, Woolsey TA (1997) Activation of a wide-spread network of inhibitory neurons in barrel cortex. Somatosens Mot Res 14:138–147

McCormick DA, Connors BW, Lighthall JW, Prince DA (1985) Comparative electrophysiology of pyramidal and sparsely spiny stellate neurons of the neocortex. J Neurophysiol 54:782–806

McCormick DA, Prince DA (1987) Post-natal development of electrophysiological properties of rat cerebral cortical pyramidal neurones. J Physiol 393:743–762

Micheva KD, Beaulieu C (1995) Neonatal sensory deprivation induces selective changes in the quantitative distribution of GABA-immunoreactive neurons in the rat barrel field cortex. J Comp Neurol 361:574–584

Micheva KD, Beaulieu C (1997) Development and plasticity of the inhibitory neocortical circuitry with an emphasis on the rodent barrel field cortex: a review. Can J Physiol Pharmacol 75:470–478

Montoro RJ, Yuste R (2004) Gap junctions in developing neocortex: a review. Brain Res Brain Res Rev 47:216–226

Moore CI, Nelson SB (1998) Spatio-temporal subthreshold receptive fields in the vibrissa representation of rat primary somatosensory cortex. J Neurophysiol 80:2882–2892

Nadarajah B, Parnavelas JG (1999) Gap junction-mediated communication in the developing and adult cerebral cortex. Novartis Found Symp 219:157–170

Nadarajah B, Thomaidou D, Evans WH, Parnavelas JG (1996) Gap junctions in the adult cerebral cortex: regional differences in their distribution and cellular expression of connexins. J Comp Neurol 376:326–342

Owens DF, Liu XL, Kriegstein AR (1999) Changing properties of $GABA_A$ receptor-mediated signaling during early neocortical development. J Neurophysiol 82:570–583

Pallas SL, Wenner P, Gonzalez-Islas C, Fagiolini M, Razak KA, Kim G, Sanes D, Roerig B (2006) Developmental plasticity of inhibitory circuitry. J Neurosci 26:10358–10361

Porter JT, Johnson CK, Agmon A (2001) Diverse types of interneurons generate thalamus-evoked feedforward inhibition in the mouse barrel cortex. J Neurosci 21:2699–2710

Prince DA (1999) Epileptogenic neurons and circuits. Adv Neurol 79:665–684

Quairiaux C, Armstrong-James M, Welker E (2007) Modified sensory processing in the barrel cortex of the adult mouse after chronic whisker stimulation. J Neurophysiol 97:2130–2147

Razak KA, Pallas SL (2006) Dark rearing reveals the mechanism underlying stimulus size tuning of superior colliculus neurons. Vis Neurosci 23:741–748

Rema V, Ebner FF (1996) Postnatal changes in NMDAR1 subunit expression in the rat trigeminal pathway to barrel field cortex. J Comp Neurol 368:165–184

Rice FL, Van der LH (1977) Development of the barrels and barrel field in the somatosensory cortex of the mouse. J Comp Neurol 171:545–560

Rivera C, Voipio J, Payne JA, Ruusuvuori E, Lahtinen H, Lamsa K, Pirvola U, Saarma M, Kaila K (1999) The K^+/Cl^- co-transporter KCC2 renders GABA hyperpolarizing during neuronal maturation. Nature 397:251–255

Shoykhet M, Land PW, Simons DJ (2005) Whisker trimming begun at birth or on postnatal day 12 affects excitatory and inhibitory receptive fields of layer IV barrel neurons. J Neurophysiol 94:3987–3995

Silberberg G, Bethge M, Markram H, Pawelzik K, Tsodyks M (2004) Dynamics of population rate codes in ensembles of neocortical neurons. J Neurophysiol 91:704–709

Simons DJ, Land PW (1987) Early experience of tactile stimulation influences organization of somatic sensory cortex. Nature 326:694–697

Spires TL, Molnar Z, Kind PC, Cordery PM, Upton AL, Blakemore C, Hannan AJ (2005) Activity-dependent regulation of synapse and dendritic spine morphology in developing barrel cortex requires phospholipase C-beta1 signalling. Cereb Cortex 15:385–393

Staiger JF, Masanneck C, Bisler S, Schleicher A, Zuschratter W, Zilles K (2002) Excitatory and inhibitory neurons express c-Fos in barrel-related columns after exploration of a novel environment. Neuroscience 109:687–699

Stryker MP (1978) Postnatal development of ocular dominance columns in layer IV of the cat's visual cortex and the effects of monocular deprivation. Arch Ital Biol 116:420–426

Sun QQ, Huguenard JR, Prince DA (2006) Barrel cortex microcircuits: thalamocortical feedforward inhibition in spiny stellate cells is mediated by a small number of fast-spiking interneurons. J Neurosci 26:1219–1230

Sun QQ, Zhang Z, Jiao Y, Zhang C, Szabo G, Erdelyi F (2009) Differential metabotropic glutamate receptor expression and modulation in two neocortical inhibitory networks. J Neurophysiol 101:2679–2692

Swadlow HA (2002) Thalamocortical control of feed-forward inhibition in awake somatosensory 'barrel' cortex. Philos Trans R Soc Lond B Biol Sci 357:1717–1727

Swadlow HA (2003) Fast-spike interneurons and feedforward inhibition in awake sensory neocortex. Cereb Cortex 13:25–32

Varju P, Katarova Z, Madarasz E, Szabo G (2001) GABA signalling during development: new data and old questions. Cell Tissue Res 305:239–246

Wang Y, Gupta A, Toledo-Rodriguez M, Wu CZ, Markram H (2002) Anatomical, physiological, molecular and circuit properties of nest basket cells in the developing somatosensory cortex. Cereb Cortex 12:395–410

Wang Y, Toledo-Rodriguez M, Gupta A, Wu C, Silberberg G, Luo J, Markram H (2004) Anatomical, physiological and molecular properties of Martinotti cells in the somatosensory cortex of the juvenile rat. J Physiol 561:65–90

West AE, Griffith EC, Greenberg ME (2002) Regulation of transcription factors by neuronal activity. Nat Rev Neurosci 3:921–931

Wiesel TN, Hubel DH (1974) Ordered arrangement of orientation columns in monkeys lacking visual experience. J Comp Neurol 158:307–318

Wilent WB, Contreras D (2005) Dynamics of excitation and inhibition underlying stimulus selectivity in rat somatosensory cortex. Nat Neurosci 8:1364–1370

Woo NH, Lu B (2006) Regulation of cortical interneurons by neurotrophins: from development to cognitive disorders. Neuroscientist 12:43–56

Woolsey TA, Van der LH (1970) The structural organization of layer IV in the somatosensory region (SI) of mouse cerebral cortex. The description of a cortical field composed of discrete cytoarchitectonic units. Brain Res 17:205–242

Xiang Z, Huguenard JR, Prince DA (1998) Cholinergic switching within neocortical inhibitory networks. Science 281:985–988

Part III
Synaptic Mechanisms

Chapter 7
GABAergic Transmission and Neuronal Network Events During Hippocampal Development

Sampsa T. Sipilä and Kai K. Kaila

7.1 Introduction

Until the middle of the last century, serious doubts were frequently expressed concerning the capability of the central nervous system (CNS) to produce endogenously generated patterns of activity, such as those postulated by the early Darwinian student of animal physiology and behavior, Thomas Henry Huxley. In his delightful book on the crayfish as a model organism in biology, Huxley (1891) writes

> If the nervous system were a mere bundle of nerve fibres extending between sensory organs and muscles, every muscular contraction would require the stimulation of that special point of the surface on which the appropriate sensory nerve ended. The contraction of several muscles at the same time, that is, the combination of movements towards one end, would be possible only if the appropriate nerves would be stimulated in the proper order, and every movement would be the direct result of external changes. The organism would be like a piano, which may be made to give out the most complicated harmonies, but is dependent on the depression of a separate key for every note that is sounded. But it is obvious that the crayfish needs no such separate impulses for the performance of highly complicated actions. ... To carry the analogy of the musical instrument further, striking a single key gives rise, not to a single note, but to a more or less elaborate tune; as if the hammer struck not a single string, but pressed down the stop of a musical box

Huxley's musical box obviously produces "motor tunes" (or "motor melodies"), which in current terminology would be synonymous to central pattern generators that produce motor programs. However, endogenous ("self-organized") activity with a core structure that is shaped without a role for momentary sensory input was viewed as something mysterious, and the mainstream behaviorist Zeitgeist sought to explain all kinds of CNS activity (not only movement related) as "chain reflexes" (Clower 1998), which corresponds to Huxley's piano. However, subsequent empirical

S.T. Sipilä
Department of Clinical Neurophysiology, Oulu University Hospital, FIN 90029, Oulu, Finland

K.K. Kaila (✉)
Department of Biosciences and Neuroscience Center, University of Helsinki,
POB 65FIN-00014, Helsinki, Finland
e-mail: Kai.Kaila@Helsinki.Fi

work carried out on the motor systems of invertebrates and vertebrates provided unequivocal evidence to support the conclusion that central pattern generators do exist (Hinde 1970; von Holst 1935, 1954; Wiersma and Ikeda 1964; Grillner 2006; Grillner and Zangger 1975). Later on, the concept of endogenous or spontaneous activity has conquered a much wider scope, and so-called "self-organized" patterns of activity have been extensively described in both the developing and adult CNS (Grillner 2006; Buzsaki 2006; Vanhatalo and Kaila 2006). Here, the term self-organized includes complex events generated and shaped by a neuronal network in the absence of any input, or in response to some information-poor stimulus (e.g., a brief phasic "trigger" input; to a tonic temporally non-patterned input; or to a transient relief from inhibition).

Much of the experimental work on endogenous pattern generation relied on experiments where incoming sensory input was blocked by deafferentation (Hamburger 1963; von Holst 1935, 1954). In this respect, demonstrating the presence of endogenous activity is more straightforward in the immature than in the mature CNS, especially if conclusive experiments are to be done in vivo. This is because complex patterns of activity are seen in immature neuronal networks at a stage of development where no sensory input is available. For example, mammalian ganglion cells fire periodic bursts of action potentials several weeks before eye opening and before maturation of the photoreceptors (Galli and Maffei 1988; Meister et al. 1991). Recent work based on full-band EEG has demonstrated the presence of endogenous intermittent activity in the immature human cortex that is particularly salient in preterm babies, and which largely disappears by the time of full-term birth (Vanhatalo and Kaila 2006; Vanhatalo et al. 2002).

Characteristically, early network activity is highly discontinuous, consisting of discrete events (e.g., retinal waves) rather than the ongoing oscillatory activity that is typical for the adult brain (Feller 1999; Ben-Ari 2001). At the cellular level, the network events are based on spatially and temporally correlated bursts of activity, and hence they have been postulated to play a key role in the formation of neuronal circuits by reinforcing connections among coactive cells (Galli and Maffei 1988; Katz and Shatz 1996). The development of GABAergic transmission is generally thought to have a crucial influence on these Hebbian mechanisms (Hensch 2005; Kanold and Shatz 2006; Katz and Crowley 2002; Zhou and Poo 2004). Notably, a tight cross-talk between GABAergic transmission and activity-dependent release of trophic factors, such as brain-derived neurotrophic factor (BDNF; see below), is likely to take place in developing hippocampal circuits (Mohajerani et al. 2007; Mohajerani and Cherubini 2006; Marty et al. 2000; Rivera et al. 2005). Early network events may also trigger endocannabinoid synthesis via elevation of intracellular Ca^{2+} (Bernard et al. 2005; Freund et al. 2003; Leinekugel et al. 1997), thereby providing a feedback mechanism to control the gross excitability of immature cortical structures (cf. Bernard et al. 2005).

In the immature hippocampus, spontaneous network events are seen in in vitro slice preparations and termed Giant Depolarizing Potentials (GDPs) (Ben-Ari et al. 1989). However, endogenous, correlated bursts of activity are not restricted to the immature hippocampus. A major class of discrete events, the Sharp Positive

Waves (SPWs), are seen throughout life in rodents, and a similar situation may hold for other mammals, including humans (Staba et al. 2004; Skaggs et al. 2007; Ulanovsky and Moss 2007; Buzsaki 1986; Buzsaki et al. 2003; Freemon and Walter 1970; Freemon et al. 1969). In the adult, these events may have several roles, including functions related to learning and memory as well as to the maintenance of neuronal circuitry. We will devote most of the present chapter to a discussion of the cellular, synaptic, and network mechanisms that generate SPWs in vivo (Leinekugel et al. 2002) as well as their putative in vitro counterparts, the GDPs, during the development of the hippocampus. The question of whether developing networks play mechanistically similar Huxleyan tunes will be briefly addressed (cf. Ben-Ari 2001).

7.2 GABAergic Transmission in the Immature Hippocampus

In light of the available information, it is clear that GABAergic mechanisms play a key role in the ontogeny of hippocampal network functions. However, the cause–effect relationships here are highly bidirectional, because various properties of GABAergic signaling themselves undergo dramatic, qualitative changes during development. A specific point that has received a large amount of attention is the "ontogenetic shift" in $GABA_A$ receptor ($GABA_A R$)-mediated transmission; in immature neurons, $GABA_A R$-mediated responses are depolarizing and sometimes even excitatory, and a shift in the reversal potential of $GABA_A R$ responses (E_{GABA}) toward more negative values takes place during neuronal maturation. The time window of this developmental shift is both neuron-type and species-specific (Kaila et al. 2008; Rivera et al. 1999; Blaesse et al. 2009), and the underlying ion-regulatory mechanisms will be reviewed below. First, however, we will describe some of the basic mechanisms and consequences of $GABA_A R$ actions in immature hippocampal and neocortical neurons.

7.2.1 Tonic Actions of GABA

Prior to the formation of functional GABAergic synapses, a tonic $GABA_A R$ conductance is present in cortical neurons (Serafini et al. 1995; LoTurco et al. 1995; Owens et al. 1999; Demarque et al. 2002). The cellular sources of the interstitial GABA that activates the tonic conductance under physiological conditions appear to be heterogenous, and include axonal growth cones that release GABA in a vesicular manner (Gao and van den Pol 2000). GABA released by astrocytes has also been shown to activate $GABA_A$ receptors at least in cultures of embryonic rat hippocampal neurons (Liu et al. 2000). Another potential source of GABA is non-vesicular release via reversal of the GABA transporters, GATs (Richerson and Wu 2003; Wu et al. 2007).

Much of the present review deals with network events that take place during the early postnatal period in rats. During this developmental stage, a pronounced tonic $GABA_A$ current persists in immature cortical pyramidal neurons, even under conditions where neuronal vesicular release is strongly suppressed (Demarque et al. 2002; Sipilä et al. 2007; Valeyev et al. 1993). Notably, pharmacological inhibition of GAT-1 leads to an increase in the magnitude of the tonic $GABA_A$ conductance. It also prolongs the decay of the slow GABAergic current component associated with the GDPs (see above) in rat hippocampal neurons (Sipilä et al. 2004, 2007). These findings indicate that GABA transport is functional and already operates in net uptake mode by birth.

7.2.2 Trophic Actions of GABA

In immature neurons, depolarizing GABAergic signaling promotes action potential activity, opening of voltage-gated Ca^{2+} channels, and activation of NMDA receptors (Ben-Ari 2002; Yuste and Katz 1991; Fukuda et al. 1998). The consequent transient elevations of the intracellular Ca^{2+} level lead to activation of a wide spectrum of signaling cascades that control various aspects of neuronal maturation and differentiation including DNA synthesis, migration, morphological maturation of individual neurons, and synaptogenesis (Wang and Kriegstein 2009). BDNF has been ascribed a key role in the trophic actions of GABA (Marty et al. 2000). However, much of the available data has been obtained in vitro, and their significance for normal neuronal development in vivo has remained somewhat unclear. Rather surprisingly, synaptogenesis and early brain development are hardly affected in knockout (KO) mice where GABA synthesis, vesicular transport, or vesicular release are eliminated (Ji et al. 1999; Wojcik et al. 2006; Verhage et al. 2000; Varoqueaux et al. 2002). Clearly, more in vivo work is needed in order to solve these discrepancies, and to elucidate the specific effects of early network events on the maturation of hippocampal neurons and networks.

7.2.3 Ion Transport and the Control of E_{GABA} in Hippocampal Neurons

Neuronal plasma membranes are equipped with a variety of ion transporters, and several of them are involved in the translocation of anions, thereby affecting E_{GABA}. These transporters have been described in recent reviews (Farrant and Kaila 2007; Blaesse et al. 2009) and hence, we will restrict the present discussion to two cation-chloride transporters, NKCC1 and KCC2, which are the major players in the developmental shift of the action of GABA in cortical neurons.

7.2.3.1 Uptake of Chloride: NKCC1

Immature neurons typically have a high internal Cl⁻ concentration maintained by specific uptake mechanisms. This leads to a rather positive value of E_{GABA} and to the depolarizing and sometimes even excitatory actions of GABA, some of which are described above. While the identity of these transporters is not clear in a wide range of neurons including the auditory brainstem and the retina (Balakrishnan et al. 2003; Vardi et al. 2000; Zhang et al. 2007), there is substantial evidence that in immature hippocampal and neocortical neurons, Cl⁻ uptake is mediated by the NKCC1 isoform of Na–K–2Cl cotransporters (Blaesse et al. 2009). Both NKCC1 and the Cl⁻ extruding K–Cl cotransporter KCC2 (see below) are secondary active transporters; they do not directly consume ATP, but take the energy for Cl⁻ uptake and extrusion from the Na⁺ and K⁺ gradients, respectively, generated and maintained by the ubiquitous Na–K ATPase.

Depolarizing GABA actions in hippocampal neurons are blocked by bumetanide (Sipilä et al. 2006b), a drug that at low concentrations (1–10 μM) selectively blocks NKCCs (Isenring et al. 1998; Payne et al. 2003). Interestingly, NKCC1 knockout mice are viable (Delpire et al. 1999; Flagella et al. 1999; Pace et al. 2000) and they do not have a conspicuous brain phenotype. Their major problems at the level of behavior seem to arise from the non-functional inner ear.

7.2.3.2 Extrusion of Chloride: KCC2

In adult CA3 and CA1 pyramidal neurons, $GABA_A$-mediated transmission is hyperpolarizing, and the extrusion of Cl⁻ needed to achieve an E_{GABA} that is more negative than the resting membrane potential (V_m) is attributable to Cl⁻ extrusion by the K–Cl cotransporter KCC2. KCC2 has not been detected in any other cells apart from central neurons, and even among mature CNS neurons some do not express this transporter (Blaesse et al. 2009).

Recent work has shown that KCC2 is expressed as two splice variants, KCC2a and KCC2b (Uvarov et al. 2007). Disruption of the KCC2-coding gene, *Slc12A5*, inhibits KCC2 expression completely and results in mice that die immediately after birth due to severe motor defects, including respiratory failure (Hubner et al. 2001). In another transgenic mouse, exon 1 of the known *Slc12A5* sequence was targeted, which was originally thought to produce a full knockout (Woo et al. 2002). For reasons that were initially unclear, 5–8% of KCC2 expression was retained. In contrast to the full knockout by Hubner et al (2001), these mice are viable after birth, but they show pronounced generalized seizures and die at an age of about 2 weeks (Woo et al. 2002). It has now become apparent that the residual KCC2 expression represents the KCC2a isoform which contains, compared to the previously described KCC2b, an alternative exon 1 (Uvarov et al. 2007). KCC2a is expressed in the neonatal brainstem and spinal cord at a level similar to KCC2b and appears to be important for some of the basic functions of these structures. In cortical neurons, KCC2b is the dominant isoform (Uvarov et al. 2007). Notably, KCC2b is responsible for the "developmental shift," as can be concluded from previous data on auditory brainstem and cortical neurons (Balakrishnan et al. 2003; Zhu et al. 2005) from mice that are now known to be KCC2b KOs (Uvarov et al. 2007).

In the present chapter, we will use "KCC2" as the term that refers to the main K–Cl cotransporter in cortical neurons, because there are no data available that would enable one to dissect the actions of the two isoforms. In addition to this, yet another neuronal K–Cl cotransporter (KCC3) has been identified in the hippocampus (Boettger et al. 2003). However, there is little information on the roles of KCC3 in the development and function of cortical neurons.

7.2.3.3 Bicarbonate and E_{GABA}

In addition to Cl$^-$, HCO$_3^-$ ions are physiologically relevant carriers of current across GABA$_A$ receptors (Kaila and Voipio 1987; Kaila 1994). The quantitative influence of HCO$_3^-$ on E_{GABA} can be readily estimated using the Goldman–Hodgkin–Katz voltage equation (Kaila 1994; Farrant and Kaila 2007). As a rule of thumb, the intracellular concentration of HCO$_3^-$ in neurons (about 15 mM at a pH of 7.1–7.2) has an influence on E_{GABA} that is equal to about 3–5 mM Cl$^-$. Hence, the depolarizing influence of HCO$_3^-$ on E_{GABA} is significant in neurons with a low internal Cl$^-$ concentration such as adult cortical neurons (Kaila et al. 1993), but it can be largely ignored in immature neurons because of their relatively high intracellular chloride levels.

7.3 Ontogeny of Hippocampal Network Events

Distinct types of network rhythms are seen in extracellular field potential recordings in the adult rodent hippocampus depending on the behavioral state of the animal. During exploratory behavior and REM sleep, the most prominent network rhythm is the theta oscillation (see Buzsaki 2006) that exerts a modulatory action on the faster gamma rhythm (Soltesz and Deschênes 1993). On the other hand, during immobile wakefulness, consummatory behaviors (such as feeding and drinking), and slow-wave sleep, a more irregular pattern is observed (Buzsaki et al. 1983). This irregular pattern contains SPWs that are associated with "ripples" (~140–200 Hz) (O'Keefe and Nadel 1978). SPWs are thought to be generated endogenously within the hippocampus by the interconnected network of CA3 pyramidal neurons (Buzsaki 1986). The SPW seems to have similar characteristics across different (perhaps all) mammalian species (Staba et al. 2004; Skaggs et al. 2007; Ulanovsky and Moss 2007; Buzsaki 1986; Buzsaki et al. 2003; Freemon and Walter 1970; Freemon et al. 1969). This is intriguing, given the fact that SPWs can be detected in both neonatal and adult rodent hippocampus (Buhl and Buzsaki 2005;

Karlsson and Blumberg 2003; Leinekugel et al. 2002; Mohns et al. 2007; Sipilä et al. 2006b), which implies a cortical time span that in the human would correspond to one which covers the last trimester of gestation and lasts for the entire life (Avishai-Eliner et al. 2002; Clancy et al. 2001). As already noted above, the presence of SPWs during such a wide time window suggests that they have several functions. In addition to their likely role in the development of neuronal circuits, SPWs are generally thought to be involved in learning and memory in the adult, especially in the transfer of hippocampally acquired information to the neocortex (Buzsaki 1989).

SPWs are the first large-scale network pattern that is seen during hippocampal development in vivo as studied in rats (Leinekugel et al. 2002; Buhl and Buzsaki 2005; Karlsson and Blumberg 2003; Mohns et al. 2007; Sipilä et al. 2006b; Leblanc and Bland 1979). Thereafter, within the first three postnatal weeks, adult-like theta and gamma oscillations emerge (Leblanc and Bland 1979; Karlsson and Blumberg 2003; Mohns et al. 2007; Lahtinen et al. 2002). SPWs are often associated with a "tail" event consisting of multi-unit bursts (Leinekugel et al. 2002), and during development, they become associated with ripples (see above).

The cellular and synaptic mechanisms generating early network rhythms have been extensively studied under in vitro conditions, mainly using hippocampal slice preparations (Ben-Ari 2001). During cortical ontogeny, large-scale population activity is preceded by local events detected as intracellular Ca^{2+} transients that involve only a few neurons as shown in embryonic and neonatal mice (Dupont et al. 2006; Crepel et al. 2007; Yuste et al. 1992). In the hippocampus, these events have been termed synchronous plateau assemblies (SPAs), that take place in the absence of chemical synaptic transmission pointing to a role of gap junctions and intrinsic membrane currents (Crepel et al. 2007). SPAs were reported to disappear at the time when GDPs emerge, and it was proposed that a transient, oxytocin-mediated shift from depolarizing to hyperpolarizing action of GABA occurring at birth promotes the emergence of SPA activity in mice. However, this is unlikely to be a common characteristic across different mammalian species, as GDPs are already seen in fetal monkey hippocampal slices (Khazipov et al. 2001).

GDPs were first described by Ben-Ari et al. (1989) at the cellular level in work on hippocampal slices from neonatal rats, where GDPs disappear by the end of the second postnatal week (Khazipov et al. 2004a; Ben-Ari et al. 1989). The temporal correlation of the disappearance of GDPs in rat slices with the development of hyperpolarizing inhibition (the ontogenetic shift in GABA action) has been one of the cornerstones of the widespread hypothesis that depolarizing GABAergic activity "sets the tune" not only for GDPs, but also for other endogenous events in the immature central nervous system (Ben-Ari et al. 2004; Ben-Ari et al. 2007). More recently, however, other groups have seen network events (termed in vitro sharp waves; Maier et al. 2003; Kubota et al. 2003; Wu et al. 2005; Foffani et al. 2007) in the mature hippocampus that share many characteristics with GDPs. Hence, it has become clear that spontaneous intermittent network events are present in postnatal slices of all ages including those from adult rats and mice, and that they are particularly prominent in the latter species. A conclusion that requires the least number of ad hoc assumptions is that GDPs are the in vitro counterparts of in vivo SPWs. The evidence for this conclusion is reviewed below.

7.4 Characteristics of "Giant Depolarizing Potentials" in the Rat Hippocampus In Vitro

"Giant Depolarizing Potentials" are named so because these spontaneous events were originally detected in intracellular recordings in the neonatal rat hippocampal CA3 pyramidal neurons in slice preparations (Ben-Ari et al. 1989, 2007). A major component of the intracellular voltage signal had a rather positive reversal potential, which resulted in a large depolarization – hence the attribute "giant." In this pioneering work, lots of attention was paid to the fact that the reversal potential of the slow depolarizing phase of the intracellular GDP was similar to that of voltage responses elicited by exogenous $GABA_A$ agonists. Furthermore, the visually dominant depolarizing component of intracellularly recorded GDPs was blocked by $GABA_A$ receptor antagonists. As already mentioned above, GDPs were found to disappear gradually during ontogeny in parallel with the maturation of hyperpolarizing GABAergic transmission that takes place during the first two postnatal weeks in the rat. On the basis of these observations, it was straightforward to conclude that GDPs are GABAergic events, i.e., network events paced by a phasic, excitatory action of GABA. In other words, the generation of GDPs and hence, their rhythmicity, was thought to be set by a synchronous excitatory action of the GABAergic interneuronal network.

As is evident from above, the acronym GDP is ambiguous; it refers both to a single-cell response that can be recorded during a network event, and also to the network event itself. This problem has been discussed elsewhere (Sipilä et al. 2005), and the context where the term GDP is used below should make it clear whether we refer to these events at the single-cell or network level.

In slices, the CA3 is considered the GDP "pacemaker region" (Ben-Ari 2001), but various other subregions of the hippocampus (CA1 and the dentate gyrus) can generate GDP-like network events in isolation (Khazipov et al. 1997; Garaschuk et al. 1998; Menendez de la Prida et al. 1998; Bolea et al. 2006). In intracellular voltage-clamp recordings in CA3 pyramidal neurons, a burst of ionotropic GABAergic and glutamatergic currents is seen during GDPs (Lamsa et al. 2000; Leinekugel et al. 1998, 2002), which are readily detected in parallel field potential recordings as a slow negative shift that is often associated with a burst of spikes. The majority of CA3 pyramidal cell spikes are confined to a 500-ms time window around the peak field potential deflection, whereas the GABAergic burst has a somewhat longer time course (Sipilä et al. 2005). Typically, GDPs occur at irregular intervals lasting from seconds to minutes. They are each followed by a refractory period of ~2–3 s, during which very little unit spike activity is seen (Sipilä et al. 2005, 2006a). GDPs are also generated by the whole-hippocampus preparation

(Leinekugel et al. 1998), where the septal pole has the highest propensity for triggering the events and acts as the pacemaker region along the longitudinal axis of the hippocampus.

7.5 Synaptic and Cellular Mechanisms Underlying GDP Generation

In the analysis of the roles of glutamatergic and GABAergic mechanisms in the generation of GDPs, we will first focus on GABA.

7.5.1 *GDPs and the Developmental Shift in GABA Action in Rat Hippocampal Slices*

The developmental expression of KCC2 in the rat hippocampus starts around birth, and an adult-like expression pattern is seen by the end of the second postnatal week. However, the temporal link between the expression of KCC2 protein and functional K–Cl extrusion that is evident in native cortical neurons (Rivera et al. 1999; Lu et al. 1999; Yamada et al. 2004) cannot be generalized to other types of neurons. For example, cultured cortical neurons show abundant KCC2 protein expression levels well in advance of the functional activation of the transporter (Khirug et al. 2005) and, interestingly, a similar situation holds for native auditory brainstem neurons (Blaesse et al. 2006). At the moment, there is only limited information available on the steps (e.g., trafficking and kinetic activation of the membrane-bound transporter) that are required for functional activation of KCC2 (Blaesse et al. 2009).

Quite unexpectedly, recent work has uncovered a role for KCC2 in synaptic transmission that is unrelated to its K–Cl cotransport activity (Li et al. 2007). A high level of KCC2 was detected in the spines of cortical neurons (Gulyas et al. 2001). In view of the role of KCC2 in GABAergic transmission, this was a rather surprising observation. However, subsequent experiments on primary cultures showed that KCC2 has a structural role in spine formation (Li et al. 2007). These results suggest that the expression of KCC2 synchronizes the development of GABAergic and glutamatergic transmission in cortical networks.

The latter part of the second postnatal week appears to be the key time point when qualitative changes in postsynaptic GABAergic responses occur in the developing rat hippocampus (Ben-Ari et al. 1989; Khazipov et al. 2004a; Tyzio et al. 2007; Rivera et al. 1999), although it should be kept in mind that there is marked heterogeneity in the actions of GABA at the level of individual neurons (Duebel et al. 2006; Szabadics et al. 2006; Khirug et al. 2008). The changes in the postsynaptic actions of GABA have been studied with various methods to examine E_{GABA}

and the driving force of GABAergic currents, or GABA's effect on the probability of spike generation (i.e., whether GABA is inhibitory or excitatory). In the current literature on the developmental shift of E_{GABA} and its consequences, however, it is often erroneously stated that depolarizing GABA actions imply excitation, and that a necessary condition for a genuinely inhibitory GABA action is an E_{GABA} that is more negative than resting V_m, i.e., hyperpolarizing. This topic has recently been discussed elsewhere (Kaila et al. 2008; Blaesse et al. 2009), but a summary is provided here:

1. Regardless of the value of E_{GABA}, the opening of the anion channels of GABA$_A$Rs will have a shunting action. Hence, moderately depolarizing GABA actions can be functionally inhibitory, and even more effective as hyperpolarizing responses, because the intrinsic $I-V$ relationship of GABA$_A$ currents shows outward rectification. In addition, even a small depolarization can lead to substantial inactivation of Na^+ channels and activation of K^+ channels. For example, adult dentate granule cells have depolarizing and strongly inhibitory GABA responses (Staley et al. 1992).
2. GABAergic transmission is not necessarily excitatory even if E_{GABA} would be more positive than what is observed as the threshold of spiking in standard somatic intracellular recordings. This is because the spike voltage threshold is not a fixed parameter, but depends on the rate of change of the membrane potential, and also on the background conductance.

The gramicidin-perforated patch clamp technique (Kyrozis and Reichling 1995) is often thought to be an ideal technique to study neuronal Cl^- extrusion, because in these measurements, intracellular Cl^- concentration is not affected by the pipette filling solution. However, measuring E_{GABA} with this (or any other) technique (Tyzio et al. 2006) in resting neurons can, at best, verify the presence of Cl^- extrusion. Notably, even a very inefficient Cl^- extrusion mechanism can be sufficient to maintain a hyperpolarizing E_{GABA} in a resting slice preparation. In the intact brain in which neurons are involved in ongoing activity and varying chloride loads, *it is the capacity of neuronal Cl^- extrusion* rather than the steady-state E_{GABA} that has a direct impact on the efficacy of inhibition. When assessing the capacity of a neuron to maintain $[Cl^-]_i$ in a range that provides a basis for inhibitory GABA action (i.e., reduces excitability), a defined Cl^- load is imposed on a cell. From such data it is possible to obtain a physiologically valid estimate of the efficacy of chloride regulation (Khirug et al. 2005; Rivera et al. 2004; Jarolimek et al. 1999). A technical point worth to emphasize here is that in experiments on E_{GABA} or on Cl^- extrusion, recording electrodes filled with Cs^+ must be avoided, because Cs^+ is a very poor substrate for KCC2 (Williams and Payne 2004) and blocks K–Cl cotransport in mammalian neurons (Thompson and Gahwiler 1989).

As emphasized elsewhere, the actions of GABA are context-dependent (Farrant and Kaila 2007; Buzsaki et al. 2007) and, notably, GABA can have "dual" actions (both excitatory and inhibitory) in an individual neuron. In a modeling study on the effects of depolarizing IPSPs (dIPSPs), Jean-Xavier et al. (2007) showed that dIPSPs were able to facilitate spike triggering by subthreshold excitatory events in their late phase.

This is because the depolarization outlasts the local, shunting conductance increase which is associated with a dIPSP. Furthermore, the time window for the enhancement of excitability by dIPSPs became wider as E_{Cl} was more depolarized, and the pro-excitatory effects started earlier when the site of dIPSP generation was further away from the excitatory input. It is obvious from these and many other observations that a dichotomous depolarizing-to-hyperpolarizing "switch" that would control the excitatory vs. inhibitory postsynaptic actions during the development of GABAergic transmission (see below) is a profound oversimplification.

7.5.2 Glutamatergic Transmission and GDPs

As stated above, the pacemaker region for GDPs in rat hippocampal slice preparations is area CA3. One of the most characteristic properties of the CA3 area is its network of glutamatergic pyramidal neurons which shows an unusually high level of interconnectivity via excitatory collaterals (Lebovitz et al. 1971; MacVicar and Dudek 1980). This structural property is likely to be a key feature for the propensity of the CA3 area to generate various types of network events, ranging from GDPs to SPWs and to interictal events. Notably, all of them show a high sensitivity to AMPA antagonists (Cohen et al. 2002; Bolea et al. 1999; Wu et al. 2005).

During the perinatal period in the rat hippocampus, development of functional GABAergic synapses occurs prior to that of glutamatergic synapses. Indeed, a vast majority of hippocampal neurons express no functional glutamatergic synapses around birth (Danglot et al. 2006; Hennou et al. 2002; Tyzio et al. 1999). This has often been taken as indirect evidence for the view that network events in early hippocampal development are driven by interneurons. However, at the network level, competitive AMPA-receptor antagonists (CNQX, NBQX, DNQX, etc) strongly inhibit GDP occurrence (Ben-Ari et al. 1989; Lamsa et al. 2000; Hollrigel et al. 1998; Bolea et al. 1999). When these drugs are combined with NMDA receptor inhibitors, GDPs are abolished (Hollrigel et al. 1998; Sipilä et al. 2005; Bolea et al. 1999; Khazipov et al. 2001). Notably, the selective AMPA blocker, GYKI 53655, completely blocks spontaneous and evoked GDPs, demonstrating a crucial role for AMPA receptor-mediated transmission in GDP generation (Bolea et al. 1999). The periodic, rhythmic activation of interneurons during GDPs is blurred into an irregular pattern by glutamatergic antagonists (Fig. 7.1), indicating that the GDP-associated interneuronal activity is a *consequence* and not a cause of pyramidal cell firing (Sipilä et al. 2005).

7.5.3 Intrinsic Bursting of CA3 Pyramidal Neurons

The belief that immature CA3 pyramidal neurons are not bursters is prevalent in the earlier literature on GDPs (e.g., Ben-Ari et al. 1989; see also Ben-Ari et al. 2007).

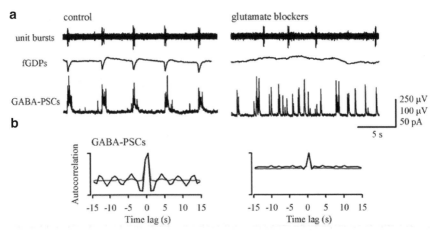

Fig. 7.1 Ionotropic glutamate-receptor mediated transmission drives patterned interneuronal activity during GDPs. (**a**) Simultaneous field potential (*upper two traces*; *top trace*: 100–600 Hz, *middle trace*: 0.05–5 Hz) and intracellular voltage-clamp recordings (*lower trace*: 0 mV, low-chloride filling solution) show that field GDPs (fGDPs) and the associated GABAergic bursts reflecting network activity of interneurons are blocked by a combined application of the glutamate receptor blockers NBQX and AP-5. Spontaneous bursting activity (unit bursts) persists in the absence of glutamatergic transmission and GDPs. (**b**) Autocorrelation histograms of spontaneous post-synaptic GABA$_A$ receptor-mediated current (GABA-PSC) intervals show that the rhythmic GABAergic activity (*left*) (at ~0.3 Hz) is abolished by the glutamate receptor antagonists (*right*). The figure is modified from Sipilä et al. (2005).

To the contrary, more recently it has been shown that immature CA3 pyramidal neurons are able to generate intrinsic bursts (Menendez de la Prida and Sanchez-Andres 2000), and that this voltage-dependent bursting takes place in the absence of synaptic transmission (Sipilä et al. 2005; Safiulina et al. 2008). The neurons are silent at negative membrane potentials, whereas a depolarization above ~ −60 mV activates a persistent Na$^+$ current that generates a slow regenerative depolarization leading to a burst of action potentials (Sipilä et al. 2006a). During a burst, Ca^{2+} enters the cell and activates a K$^+$ current that generates a slow afterhyperpolarization (sAHP), and the sAHP accounts for burst termination (Sipilä et al. 2006a). The neuron is in a relative refractory state during the sAHP (see below). Notably, while the membrane voltage is within the activation range of the hyperpolarization-activated cation current I_h during the post-burst sAHP, blocking I_h has little effect on the intrinsic pyramidal bursts (Sipilä et al. 2006a). At threshold level, the burst frequency is around 0.2 Hz and can increase up to ~1.5 Hz in response to a tonic depolarization (Sipilä et al. 2005; Menendez de la Prida and Sanchez-Andres 2000). Each burst consists of 2–8 spikes that occur at a rate of 10–50 Hz. Mature CA3 pyramidal cells also generate voltage-dependent intrinsic bursts that occur at a similar frequency range as the bursts seen in the immature neurons (Hablitz and Johnston 1981; Kandel and Spencer 1961; Kandel et al. 1961; Wong and Prince 1981). However, a notable difference between the mature vs. the immature CA3 pyramidal neurons is that the former can fire at a much higher frequency (50–400 Hz) during a burst (Kandel and Spencer 1961; Kandel et al. 1961; Ranck 1973).

7.5.4 CA3 Pyramidal Neurons as Conditional Pacemakers in GDP Generation

There is a striking similarity between the temporal patterns of single-unit CA3 pyramidal cell bursts and GDPs; both have a preferred (modal) frequency at ~0.2–0.5 Hz, and a tonic depolarization increases the frequency of both the single-cell and network events within a similar (~0.2–1.5 Hz) frequency range (Sipilä et al. 2005). These observations are consistent with the idea that the temporal pattern of GDP activity has its roots in the intrinsic properties of the CA3 pyramidal neurons. The data reviewed below provide further evidence for this view.

The actions of ionotropic GABA and glutamate receptor antagonists on the burst activity of immature CA3 pyramidal neurons are qualitatively different. A blockade of glutamate receptors desynchronizes pyramidal bursts but does not abolish them. On the other hand, $GABA_A$ receptor antagonists reduce the frequency of the synchronous events and sometimes completely block them (Sipilä et al. 2005). Whereas the frequency of field GDPs is reduced, their amplitude is typically enhanced (Lamsa et al. 2000). The inhibitory effect of $GABA_A$ receptor antagonists on the network events is readily explained by the finding that these drugs inhibit the burst activity of the pyramidal neurons. This effect takes place also in the absence of glutamatergic transmission (i.e., in the absence of GDPs), because (as explained above) the immature CA3 pyramidal neurons are subject to an endogenous, strongly depolarizing input mediated by synaptic and tonic $GABA_A$ receptors, which is blocked by $GABA_A$ receptor antagonists. It should be re-emphasized that because the interneuronal network activity in itself is temporally non-patterned, the above suppressing actions of $GABA_A$ antagonists are caused by a hyperpolarization and a consequent decrease in the overall excitability of the CA3 neurons.

The threshold for the activation of the persistent Na^+ current is ~10 mV negative compared to the threshold of action potential generation, and any depolarization that activates the persistent Na^+ current is able to promote the cellular bursts (Sipilä et al. 2005, 2006a). Taking advantage of this fact, more evidence for the conclusion that the "cellular GDP pacemakers" reside within the CA3 pyramidal neuron network was based on the observation that GDP occurrence is triggered by a pharmacologically induced tonic GABAergic depolarization in the complete absence of synaptic $GABA_A$ receptor transmission. Even more strikingly, we observed that GDPs were also promoted by a tonic depolarization imposed by elevation of extracellular K^+ concentration in the complete absence of both synaptic and tonic $GABA_A$ receptor-mediated signaling (Sipilä et al. 2005).

The relative refractory period of GDPs, defined as the minimum inter-GDP interval under standard experimental conditions, is similar to that of the intrinsic bursts of immature CA3 pyramidal neurons (Sipilä et al. 2005, 2006a; see also

Agmon and Wells 2003). In light of the present mechanistic scheme of GDP generation, this similarity is to be expected, since a large number of pyramidal neurons that are co-activated during a GDP will become simultaneously refractory due to the sAHP during the post-GDP period. Hence, they cannot contribute to a network event during this period. The refractory period is not absolute, however, because a strong enough depolarizing input can trigger both the cellular and network events. Furthermore, various blockers of the Ca^{2+}-activated K^+ current abolish the refractory period seen at the level of individual neurons and the network (Sipilä et al. 2006a). GDP inter-event intervals are often longer than the relative refractory period, which means that other, as yet unidentified mechanisms in addition to the Ca^{2+}-activated K^+ current determine the duration of the inter-GDP interval.

7.6 Conclusions

In the present chapter, our focus has been on hippocampal GDPs, because they offer an excellent opportunity to examine the validity and scope of "general rules" that are proposed to underlie the functional development of neuronal networks (Ben-Ari 2006; Ben-Ari et al. 2007). Two major properties that have been attributed to GDPs are that (1) they are present during a restricted window of development corresponding to the period when GABA has a depolarizing action and (2) interneurons are crucial in setting the pace of the events (Ben-Ari et al. 2007). In light of the available evidence reviewed above, neither of these properties is a defining characteristic for GDPs. Paradoxically, in contrast to e.g., retinal waves, it now seems that hippocampal GDPs, which were originally considered a prototype of early network events, are an exceptional type of early event in that their in vivo counterparts – the SPWs – are retained throughout life (Leinekugel et al. 2002; Sipilä et al. 2006b; Mohns et al. 2007; Buzsaki 1986). Consistent with this observation, work carried out on rodent slices shows that bursting activity of the pyramidal cell population drives hippocampal network events throughout life (Sipilä et al. 2005; Miles and Wong 1983; Traub et al. 1989). SPWs have been recorded invasively in early neonatal rats (postnatal day 2), and while the SPWs become associated with high-frequency ripples during development (Buhl and Buzsaki 2005; Mohns et al. 2007), their basic appearance remains largely unchanged throughout the postnatal life of rodents. Undoubtedly, depolarizing GABA plays a facilitatory or permissive role in the generation of both GDPs and neonatal SPWs. It would be difficult to envisage a scenario, however, where these events would initially be paced by the interneuronal network and thereafter, at some time point that has never been specified, the pacing role would be taken over by the pyramidal neuronal network.

In light of the above discussion, it is evident that depolarizing GABA actions do not set the "melody" played by the developing hippocampus. Here, a comparison between the immature hippocampus and neocortex is interesting. The dominant pattern of neocortical spontaneous activity in vitro, a sharp potential that is reminiscent of hippocampal GDP/SPWs, is sensitive to $GABA_A$ receptor antagonists

(Rheims et al. 2008) in a manner similar to hippocampal GDPs (see above). In contrast to this, the dominant pattern in vivo, the "spindle-bursts" (Khazipov et al. 2004b; Khazipov and Luhmann 2006), are not markedly affected by complete block of $GABA_AR$ transmission by a receptor antagonist or by blocking the depolarizing action of GABA with bumetanide (Minlebaev et al. 2007). The block of GABAergic transmission produces a slight increase in the occurrence of both evoked and spontaneous spindle-bursts. However, their spatial extent is increased, suggesting that $GABA_AR$ transmission exerts "surround inhibition" and thereby plays a role in the spatial compartmentalization of these events. Although GABAergic transmission contributes minimally to the pacing of the spindle-bursts, it is notable that the AMPA/kainate receptor antagonist CNQX completely eliminates the events (Minlebaev et al. 2007).

To summarize, the neonatal pyramidal CA3 neurons are cellular pacemakers, and when a sufficient number of them fire within a confined time window, a critical level of network excitation is attained and a GDP is generated in its well-known all-or-none manner. Unlike the glutamatergic network, the interneuronal network is not capable of producing robustly patterned activity by itself. Hence, it can be concluded that glutamatergic transmission plays an instructive role in hippocampal GDP generation, while GABA has a facilitatory or permissive mode of action. At a more general level, it seems that some of the broad generalizations and "rules" regarding the role of depolarizing GABA actions in the generation of endogenous activity in the developing brain need to be re-evaluated.

Acknowledgments The authors' original research work has been supported by the Academy of Finland, the Jane and Aatos Erkko Foundation and the Sigrid Jusélius Foundation. The authors thank Drs. Roustem Khazipov, Liset Menendez de la Prida, Michael O'Donovan, Eva Ruusuvuori, and Else Tolner for constructive comments on an early draft of the manuscript.

References

Agmon A, Wells JE (2003) The role of the hyperpolarization-activated cationic current I(h) in the timing of interictal bursts in the neonatal hippocampus. J Neurosci 23:3658–3668
Avishai-Eliner S, Brunson KL, Sandman CA, Baram TZ (2002) Stressed-out, or in (utero)? Trends Neurosci 25:518–524
Balakrishnan V, Becker M, Lohrke S, Nothwang HG, Guresir E, Friauf E (2003) Expression and function of chloride transporters during development of inhibitory neurotransmission in the auditory brainstem. J Neurosci 23:4134–4145
Ben-Ari Y (2001) Developing networks play a similar melody. Trends Neurosci 24:353–360
Ben-Ari Y (2002) Excitatory actions of gaba during development: the nature of the nurture. Nat Rev Neurosci 3:728–739
Ben-Ari Y (2006) Basic developmental rules and their implications for epilepsy in the immature brain. Epileptic Disord 8:91–102
Ben-Ari Y, Cherubini E, Corradetti R, Gaiarsa JL (1989) Giant synaptic potentials in immature rat CA3 hippocampal neurones. J Physiol (Lond) 416:303–325
Ben-Ari Y, Khalilov I, Represa A, Gozlan H (2004) Interneurons set the tune of developing networks. Trends Neurosci 27:422–427

Ben-Ari Y, Gaiarsa JL, Tyzio R, Khazipov R (2007) GABA: a pioneer transmitter that excites immature neurons and generates primitive oscillations. Physiol Rev 87:1215–1284

Bernard C, Milh M, Morozov YM, Ben-Ari Y, Freund TF, Gozlan H (2005) Altering cannabinoid signaling during development disrupts neuronal activity. Proc Natl Acad Sci USA 102:9388–9393

Blaesse P, Guillemin I, Schindler J, Schweizer M, Delpire E, Khiroug L, Friauf E, Nothwang HG (2006) Oligomerization of KCC2 correlates with development of inhibitory neurotransmission. J Neurosci 26:10407–10419

Blaesse P, Airaksinen MS, Rivera C, Kaila K (2009) Cation-chloride cotransporters and neuronal function. Neuron 61(6):820–838

Boettger T, Rust MB, Maier H, Seidenbecher T, Schweizer M, Keating DJ, Faulhaber J, Ehmke H, Pfeffer C, Scheel O, Lemcke B, Horst J, Leuwer R, Pape HC, Volkl H, Hubner CA, Jentsch TJ (2003) Loss of K-Cl co-transporter KCC3 causes deafness, neurodegeneration and reduced seizure threshold. EMBO J 22:5422–5434

Bolea S, Avignone E, Berretta N, Sanchez-Andres JV, Cherubini E (1999) Glutamate controls the induction of GABA-mediated giant depolarizing potentials through AMPA receptors in neonatal rat hippocampal slices. J Neurophysiol 81:2095–2102

Bolea S, Sanchez-Andres JV, Huang X, Wu JY (2006) Initiation and propagation of neuronal coactivation in the developing hippocampus. J Neurophysiol 95:552–561

Buhl DL, Buzsaki G (2005) Developmental emergence of hippocampal fast-field "ripple" oscillations in the behaving rat pups. Neuroscience 134:1423–1430

Buzsaki G (1986) Hippocampal sharp waves: their origin and significance. Brain Res 398:242–252

Buzsaki G (1989) Two-stage model of memory trace formation: a role for "noisy" brain states. Neuroscience 31:551–570

Buzsaki G (2006) Rhythms of the brain. Oxford University Press, USA

Buzsaki G, Leung LW, Vanderwolf CH (1983) Cellular bases of hippocampal EEG in the behaving rat. Brain Res 287:139–171

Buzsaki G, Buhl DL, Harris KD, Csicsvari J, Czeh B, Morozov A (2003) Hippocampal network patterns of activity in the mouse. Neuroscience 116:201–211

Buzsaki G, Kaila K, Raichle M (2007) Inhibition and brain work. Neuron 56:771–783

Clancy B, Darlington RB, Finlay BL (2001) Translating developmental time across mammalian species. Neuroscience 105:7–17

Clower WT (1998) Early contributions to the reflex chain hypothesis. J Hist Neurosci 7:32–42

Cohen I, Navarro V, Clemenceau S, Baulac M, Miles R (2002) On the origin of interictal activity in human temporal lobe epilepsy in vitro. Science 298:1418–1421

Crepel V, Aronov D, Jorquera I, Represa A, Ben-Ari Y, Cossart R (2007) A parturition-associated nonsynaptic coherent activity pattern in the developing hippocampus. Neuron 54:105–120

Danglot L, Triller A, Marty S (2006) The development of hippocampal interneurons in rodents. Hippocampus 16:1032–1060

Delpire E, Lu J, England R, Dull C, Thorne T (1999) Deafness and imbalance associated with inactivation of the secretory Na-K-2Cl co-transporter. Nat Genet 22:192–195

Demarque M, Represa A, Becq H, Khalilov I, Ben-Ari Y, Aniksztejn L (2002) Paracrine intercellular communication by a Ca2+- and SNARE-independent release of GABA and glutamate prior to synapse formation. Neuron 36:1051–1061

Duebel J, Haverkamp S, Schleich W, Feng G, Augustine GJ, Kuner T, Euler T (2006) Two-photon imaging reveals somatodendritic chloride gradient in retinal ON-type bipolar cells expressing the biosensor Clomeleon. Neuron 49:81–94

Dupont E, Hanganu IL, Kilb W, Hirsch S, Luhmann HJ (2006) Rapid developmental switch in the mechanisms driving early cortical columnar networks. Nature 439:79–83

Farrant M, Kaila K (2007) The cellular, molecular and ionic basis of GABA(A) receptor signalling. Prog Brain Res 160:59–87

Feller MB (1999) Spontaneous correlated activity in developing neural circuits. Neuron 22:653–656

Flagella M, Clarke LL, Miller ML, Erway LC, Giannella RA, Andringa A, Gawenis LR, Kramer J, Duffy JJ, Doetschman T, Lorenz JN, Yamoah EN, Cardell EL, Shull GE (1999) Mice lacking the basolateral Na-K-2Cl cotransporter have impaired epithelial chloride secretion and are profoundly deaf. J Biol Chem 274:26946–26955

Foffani G, Uzcategui YG, Gal B, Menendez de la Prida L (2007) Reduced spike-timing reliability correlates with the emergence of fast ripples in the rat epileptic hippocampus. Neuron 55:930–941

Freemon FR, Walter RD (1970) Electrical activity of human limbic system during sleep. Compr Psychiatry 11:544–551

Freemon FR, McNew JJ, Adey WR (1969) Sleep of unrestrained chimpanzee: cortical and subcortical recordings. Exp Neurol 25:129–137

Freund TF, Katona I, Piomelli D (2003) Role of endogenous cannabinoids in synaptic signaling. Physiol Rev 83:1017–1066

Fukuda A, Muramatsu K, Okabe A, Shimano Y, Hida H, Fujimoto I, Nishino H (1998) Changes in intracellular Ca2+ induced by GABAA receptor activation and reduction in Cl- gradient in neonatal rat neocortex. J Neurophysiol 79:439–446

Galli L, Maffei L (1988) Spontaneous impulse activity of rat retinal ganglion cells in prenatal life. Science 242:90–91

Gao XB, van den Pol AN (2000) GABA release from mouse axonal growth cones. J Physiol 523(Pt 3):629–637

Garaschuk O, Hanse E, Konnerth A (1998) Developmental profile and synaptic origin of early network oscillations in the CA1 region of rat neonatal hippocampus. J Physiol (Lond) 507:219–236

Grillner S (2006) Biological pattern generation: the cellular and computational logic of networks in motion. Neuron 52:751–766

Grillner S, Zangger P (1975) How detailed is the central pattern generation for locomotion? Brain Res 88:367–371

Gulyas AI, Sik A, Payne JA, Kaila K, Freund TF (2001) The KCl cotransporter, KCC2, is highly expressed in the vicinity of excitatory synapses in the rat hippocampus. Eur J Neurosci 13:2205–2217

Hablitz JJ, Johnston D (1981) Endogenous nature of spontaneous bursting in hippocampal pyramidal neurons. Cell Mol Neurobiol 1:325–334

Hamburger V (1963) Some aspects of the embryology of behavior. Q Rev Biol 38:342–365

Hennou S, Khalilov I, Diabira D, Ben-Ari Y, Gozlan H (2002) Early sequential formation of functional GABA(A) and glutamatergic synapses on CA1 interneurons of the rat foetal hippocampus. Eur J Neurosci 16:197–208

Hensch TK (2005) Critical period plasticity in local cortical circuits. Nat Rev Neurosci 6:877–888

Hinde RA (1970) Animal behaviour. A synthesis of ethology and comparative psychology. McGraw-Hill, New York

Hollrigel GS, Ross ST, Soltesz I (1998) Temporal patterns and depolarizing actions of spontaneous GABAA receptor activation in granule cells of the early postnatal dentate gyrus. J Neurophysiol 80:2340–2351

Hubner CA, Stein V, Hermans-Borgmeyer I, Meyer T, Ballanyi K, Jentsch TJ (2001) Disruption of KCC2 reveals an essential role of K-Cl cotransport already in early synaptic inhibition. Neuron 30:515–524

Huxley TH (1891) An introduction to the study of zoology, illustrated by the crayfish. D. Appleton and Co., New York

Isenring P, Jacoby SC, Payne JA, Forbush B III (1998) Comparison of Na-K-Cl cotransporters. NKCC1, NKCC2, and the HEK cell Na-L-Cl cotransporter. J Biol Chem 273:11295–11301

Jarolimek W, Lewen A, Misgeld U (1999) A furosemide-sensitive K+-Cl- cotransporter counteracts intracellular Cl- accumulation and depletion in cultured rat midbrain neurons. J Neurosci 19:4695–4704

Jean-Xavier C, Mentis GZ, O'Donovan MJ, Cattaert D, Vinay L (2007) Dual personality of GABA/glycine-mediated depolarizations in immature spinal cord. Proc Natl Acad Sci U S A 104:11477–11482

Ji F, Kanbara N, Obata K (1999) GABA and histogenesis in fetal and neonatal mouse brain lacking both the isoforms of glutamic acid decarboxylase. Neurosci Res 33:187–194

Kaila K (1994) Ionic basis of GABAA receptor channel function in the nervous system. Prog Neurobiol 42:489–537

Kaila K, Voipio J (1987) Postsynaptic fall in intracellular pH induced by GABA-activated bicarbonate conductance. Nature 330:163–165

Kaila K, Voipio J, Paalasmaa P, Pasternack M, Deisz RA (1993) The role of bicarbonate in GABAA receptor-mediated IPSPs of rat neocortical neurones. J Physiol (Lond) 464: 273–289

Kaila K, Blaesse P, Sipilä ST (2008) Development of GABAergic signaling: from molecules to emerging networks. In: Blumberg MS, Freeman JH, Robinson SR (eds) Oxford handbook of developmental behavioral neuroscience. Oxford University Press, Oxford

Kandel ER, Spencer WA (1961) Electrophysiology of hippocampal neurons. II. After-potentials and repetitive firing. J Neurophysiol 24:243–259

Kandel ER, Spencer WA, Brinley FJ Jr (1961) Electrophysiology of hippocampal neurons. I. Sequential invasion and synaptic organization. J Neurophysiol 24:225–242

Kanold PO, Shatz CJ (2006) Subplate neurons regulate maturation of cortical inhibition and outcome of ocular dominance plasticity. Neuron 51:627–638

Karlsson KA, Blumberg MS (2003) Hippocampal theta in the newborn rat is revealed under conditions that promote REM sleep. J Neurosci 23:1114–1118

Katz LC, Crowley JC (2002) Development of cortical circuits: lessons from ocular dominance columns. Nat Rev Neurosci 3:34–42

Katz LC, Shatz CJ (1996) Synaptic activity and the construction of cortical circuits. Science 274:1133–1138

Khazipov R, Luhmann HJ (2006) Early patterns of electrical activity in the developing cerebral cortex of humans and rodents. Trends Neurosci 29:414–418

Khazipov R, Leinekugel X, Khalilov I, Gaiarsa JL, Ben-Ari Y (1997) Synchronization of GABAergic interneuronal network in CA3 subfield of neonatal rat hippocampal slices. J Physiol (Lond) 498:763–772

Khazipov R, Esclapez M, Caillard O, Bernard C, Khalilov I, Tyzio R, Hirsch J, Dzhala V, Berger B, Ben-Ari Y (2001) Early development of neuronal activity in the primate hippocampus in utero. J Neurosci 21:9770–9781

Khazipov R, Khalilov I, Tyzio R, Morozova E, Ben-Ari Y, Holmes GL (2004a) Developmental changes in GABAergic actions and seizure susceptibility in the rat hippocampus. Eur J Neurosci 19:590–600

Khazipov R, Sirota A, Leinekugel X, Holmes GL, Ben-Ari Y, Buzsaki G (2004b) Early motor activity drives spindle bursts in the developing somatosensory cortex. Nature 432:758–761

Khirug S, Huttu K, Ludwig A, Smirnov S, Voipio J, Rivera C, Kaila K, Khiroug L (2005) Distinct properties of functional KCC2 expression in immature mouse hippocampal neurons in culture and in acute slices. Eur J Neurosci 21:899–904

Khirug S, Yamada J, Afzalov R, Voipio J, Khiroug L, Kaila K (2008) GABAergic depolarization of the axon initial segment in cortical principal neurons is caused by the Na-K-2Cl cotransporter NKCC1. J Neurosci 28(18):4635–4639

Kubota D, Colgin LL, Casale M, Brucher FA, Lynch G (2003) Endogenous waves in hippocampal slices. J Neurophysiol 89:81–89

Kyrozis A, Reichling DB (1995) Perforated-patch recording with gramicidin avoids artifactual changes in intracellular chloride concentration. J Neurosci Methods 57:27–35

Lahtinen H, Palva JM, Sumanen S, Voipio J, Kaila K, Taira T (2002) Postnatal development of rat hippocampal gamma rhythm in vivo. J Neurophysiol 88:1469–1474

Lamsa K, Palva JM, Ruusuvuori E, Kaila K, Taira T (2000) Synaptic GABA(A) activation inhibits AMPA-kainate receptor-mediated bursting in the newborn (P0–P2) rat hippocampus. J Neurophysiol 83:359–366

Leblanc MO, Bland BH (1979) Developmental aspects of hippocampal electrical activity and motor behavior in the rat. Exp Neurol 66:220–237

Lebovitz RM, Dichter M, Spencer WA (1971) Recurrent excitation in the CA3 region of cat hippocampus. Int J Neurosci 2:99–107

Leinekugel X, Medina I, Khalilov I, Ben-Ari Y, Khazipov R (1997) Ca2+ oscillations mediated by the synergistic excitatory actions of GABA(A) and NMDA receptors in the neonatal hippocampus. Neuron 18:243–255

Leinekugel X, Khalilov I, Ben-Ari Y, Khazipov R (1998) Giant depolarizing potentials: the septal pole of the hippocampus paces the activity of the developing intact septohippocampal complex in vitro. J Neurosci 18:6349–6357

Leinekugel X, Khazipov R, Cannon R, Hirase H, Ben-Ari Y, Buzsaki G (2002) Correlated bursts of activity in the neonatal hippocampus in vivo. Science 296:2049–2052

Li H, Khirug S, Cai C, Ludwig A, Blaesse P, Kolikova J, Afzalov R, Coleman SK, Lauri S, Airaksinen MS, Keinanen K, Khiroug L, Saarma M, Kaila K, Rivera C (2007) KCC2 interacts with the dendritic cytoskeleton to promote spine development. Neuron 56:1019–1033

Liu QY, Schaffner AE, Chang YH, Maric D, Barker JL (2000) Persistent activation of GABA(A) receptor/Cl(-) channels by astrocyte- derived GABA in cultured embryonic rat hippocampal neurons. J Neurophysiol 84:1392–1403

LoTurco JJ, Owens DF, Heath MJ, Davis MB, Kriegstein AR (1995) GABA and glutamate depolarize cortical progenitor cells and inhibit DNA synthesis. Neuron 15:1287–1298

Lu J, Karadsheh M, Delpire E (1999) Developmental regulation of the neuronal-specific isoform of K- Cl cotransporter KCC2 in postnatal rat brains. J Neurobiol 39:558–568

MacVicar BA, Dudek FE (1980) Local synaptic circuits in rat hippocampus: interactions between pyramidal cells. Brain Res 184:220–223

Maier N, Nimmrich V, Draguhn A (2003) Cellular and network mechanisms underlying spontaneous sharp wave-ripple complexes in mouse hippocampal slices. J Physiol (Lond) 550:873–887

Marty S, Wehrle R, Sotelo C (2000) Neuronal activity and brain-derived neurotrophic factor regulate the density of inhibitory synapses in organotypic slice cultures of postnatal hippocampus. J Neurosci 20:8087–8095

Meister M, Wong RO, Baylor DA, Shatz CJ (1991) Synchronous bursts of action potentials in ganglion cells of the developing mammalian retina. Science 252:939–943

Menendez de la Prida L, Sanchez-Andres JV (2000) Heterogeneous populations of cells mediate spontaneous synchronous bursting in the developing hippocampus through a frequency-dependent mechanism. Neuroscience 97:227–241

Menendez de la Prida L, Bolea S, Sanchez-Andres JV (1998) Origin of the synchronized network activity in the rabbit developing hippocampus. Eur J Neurosci 10:899–906

Miles R, Wong RK (1983) Single neurones can initiate synchronized population discharge in the hippocampus. Nature 306:371–373

Minlebaev M, Ben-Ari Y, Khazipov R (2007) Network mechanisms of spindle-burst oscillations in the neonatal rat barrel cortex in vivo. J Neurophysiol 97:692–700

Mohajerani MH, Cherubini E (2006) Role of giant depolarizing potentials in shaping synaptic currents in the developing hippocampus. Critical Rev Neurobiol 18:13–23

Mohajerani MH, Sivakumaran S, Zacchi P, Aguilera P, Cherubini E (2007) Correlated network activity enhances synaptic efficacy via BDNF and the ERK pathway at immature CA3 CA1 connections in the hippocampus. Proc Natl Acad Sci USA 104:13176–13181

Mohns EJ, Karlsson KA, Blumberg MS (2007) Developmental emergence of transient and persistent hippocampal events and oscillations and their association with infant seizure susceptibility. Eur J Neurosci 26:2719–2730

O'Keefe J, Nadel L (1978) The hippocampus as a cognitive map. Oxford University Press, Oxford

Owens DF, Liu X, Kriegstein AR (1999) Changing properties of GABA(A) receptor-mediated signaling during early neocortical development. J Neurophysiol 82:570–583

Pace AJ, Lee E, Athirakul K, Coffman TM, O'Brien DA, Koller BH (2000) Failure of spermatogenesis in mouse lines deficient in the Na(+)-K(+)-2Cl(-) cotransporter. J Clin Invest 105:441–450

Payne JA, Rivera C, Voipio J, Kaila K (2003) Cation-chloride co-transporters in neuronal communication, development and trauma. Trends Neurosci 26:199–206

Ranck JB Jr (1973) Studies on single neurons in dorsal hippocampal formation and septum in unrestrained rats. I. Behavioral correlates and firing repertoires. Exp Neurol 41:461–531

Rheims S, Minlebaev M, Ivanov A, Represa A, Khazipov R, Holmes GL, Ben Ari Y, Zilberter Y (2008) Excitatory GABA in rodent developing neocortex in vitro. J Neurophysiol 100:609–619

Richerson GB, Wu Y (2003) Dynamic equilibrium of neurotransmitter transporters: not just for reuptake anymore. J Neurophysiol 90:1363–1374

Rivera C, Voipio J, Payne JA, Ruusuvuori E, Lahtinen H, Lamsa K, Pirvola U, Saarma M, Kaila K (1999) The K+/Cl- co-transporter KCC2 renders GABA hyperpolarizing during neuronal maturation. Nature 397:251–255

Rivera C, Voipio J, Thomas-Crusells J, Li H, Emri Z, Sipilä S, Payne JA, Minichiello L, Saarma M, Kaila K (2004) Mechanism of activity-dependent downregulation of the neuron-specific K-Cl cotransporter KCC2. J Neurosci 24:4683–4691

Rivera C, Voipio J, Kaila K (2005) Two developmental switches in GABAergic signalling: the K+-Cl- cotransporter KCC2 and carbonic anhydrase CAVII. J Physiol (Lond) 562:27–36

Safiulina VF, Zacchi P, Taglialatela M, Yaari Y, Cherubini E (2008) Low expression of Kv7/M channels facilitates intrinsic and network bursting in the developing rat hippocampus. J Physiol 586:5437–5453

Serafini R, Valeyev AY, Barker JL, Poulter MO (1995) Depolarizing GABA-activated Cl- channels in embryonic rat spinal and olfactory bulb cells. J Physiol (Lond) 488:371–386

Sipilä S, Huttu K, Voipio J, Kaila K (2004) GABA uptake via GABA transporter-1 modulates GABAergic transmission in the immature hippocampus. J Neurosci 24:5877–5880

Sipilä ST, Huttu K, Soltesz I, Voipio J, Kaila K (2005) Depolarizing GABA acts on intrinsically bursting pyramidal neurons to drive giant depolarizing potentials in the immature hippocampus. J Neurosci 25:5280–5289

Sipilä ST, Huttu K, Voipio J, Kaila K (2006a) Intrinsic bursting of immature CA3 pyramidal neurons and consequent giant depolarizing potentials are driven by a persistent Na current and terminated by a slow Ca-activated K current. Eur J Neurosci 23:2330–2338

Sipilä ST, Schuchmann S, Voipio J, Yamada J, Kaila K (2006b) The cation-chloride cotransporter NKCC1 promotes sharp waves in the neonatal rat hippocampus. J Physiol 573:765–773

Sipilä ST, Voipio J, Kaila K (2007) GAT-1 acts to limit a tonic GABA(A) current in rat CA3 pyramidal neurons at birth. Eur J Neurosci 25:717–722

Skaggs WE, McNaughton BL, Permenter M, Archibeque M, Vogt J, Amaral DG, Barnes CA (2007) EEG sharp waves and sparse ensemble unit activity in the macaque hippocampus. J Neurophysiol 98:898–910

Soltesz I, Deschênes M (1993) Low- and high-frequency membrane potential oscillations during theta activity in CA1 and CA3 pyramidal neurons of the rat hippocampus under ketamine-xylazine anesthesia. J Neurophysiol 70:97–116

Staba RJ, Wilson CL, Bragin A, Jhung D, Fried I, Engel J Jr (2004) High-frequency oscillations recorded in human medial temporal lobe during sleep. Ann Neurol 56:108–115

Staley KJ, Otis TS, Mody I (1992) Membrane properties of dentate gyrus granule cells: comparison of sharp microelectrode and whole-cell recordings. J Neurophysiol 67:1346–1358

Szabadics J, Varga C, Molnar G, Olah S, Barzo P, Tamas G (2006) Excitatory effect of GABAergic axo-axonic cells in cortical microcircuits. Science 311:233–235

Thompson SM, Gahwiler BH (1989) Activity-dependent disinhibition. II. Effects of extracellular potassium, furosemide, and membrane potential on ECl- in hippocampal CA3 neurons. J Neurophysiol 61:512–523

Traub RD, Miles R, Wong RK (1989) Model of the origin of rhythmic population oscillations in the hippocampal slice. Science 243:1319–1325

Tyzio R, Represa A, Jorquera I, Ben-Ari Y, Gozlan H, Aniksztejn L (1999) The establishment of GABAergic and glutamatergic synapses on CA1 pyramidal neurons is sequential and correlates with the development of the apical dendrite. J Neurosci 19:10372–10382

Tyzio R, Cossart R, Khalilov I, Minlebaev M, Hubner CA, Represa A, Ben Ari Y, Khazipov R (2006) Maternal oxytocin triggers a transient inhibitory switch in GABA signaling in the fetal brain during delivery. Science 314:1788–1792

Tyzio R, Holmes GL, Ben-Ari Y, Khazipov R (2007) Timing of the developmental switch in GABA(A) mediated signaling from excitation to inhibition in CA3 rat hippocampus using gramicidin perforated patch and extracellular recordings. Epilepsia 48(Suppl 5):96–105

Ulanovsky N, Moss CF (2007) Hippocampal cellular and network activity in freely moving echo-locating bats. Nat Neurosci 10:224–233

Uvarov P, Ludwig A, Markkanen M, Pruunsild P, Kaila K, Delpire E, Timmusk T, Rivera C, Airaksinen MS (2007) A novel N-terminal isoform of the neuron-specific K-Cl cotransporter KCC2. J Biol Chem 282:30570–30576

Valeyev AY, Cruciani RA, Lange GD, Smallwood VS, Barker JL (1993) Cl- channels are randomly activated by continuous GABA secretion in cultured embryonic rat hippocampal neurons. Neurosci Lett 155:199–203

Vanhatalo S, Kaila K (2006) Development of neonatal EEG activity: from phenomenology to physiology. Semin Fetal Neonatal Med 11(6):471–478

Vanhatalo S, Tallgren P, Andersson S, Sainio K, Voipio J, Kaila K (2002) DC-EEG discloses prominent, very slow activity patterns during sleep in preterm infants. Clin Neurophysiol 113:1822–1825

Vardi N, Zhang LL, Payne JA, Sterling P (2000) Evidence that different cation chloride cotransporters in retinal neurons allow opposite responses to GABA. J Neurosci 20:7657–7663

Varoqueaux F, Sigler A, Rhee JS, Brose N, Enk C, Reim K, Rosenmund C (2002) Total arrest of spontaneous and evoked synaptic transmission but normal synaptogenesis in the absence of Munc13-mediated vesicle priming. Proc Natl Acad Sci USA 99:9037–9042

Verhage M, Maia AS, Plomp JJ, Brussaard AB, Heeroma JH, Vermeer H, Toonen RF, Hammer RE, van den Berg TK, Missler M, Geuze HJ, Sudhof TC (2000) Synaptic assembly of the brain in the absence of neurotransmitter secretion. Science 287:864–869

von Holst E (1935) Erregungsbildung und Erregungsleitung im Fischrückenmark. Pflugers Arch 235:345–359

von Holst E (1954) Relations between the central nervous system and the peripheral organs. Br J Anim Behav 2:89–94

Wang DD, Kriegstein AR (2009) Defining the Role of GABA in Cortical Development. J Physiol 587(Pt 9):1873–1879

Wiersma CA, Ikeda K (1964) Interneurons commanding swimmeret movements in the crayfish, procambarus clarki (girard). Comp Biochem Physiol 12:509–525

Williams JR, Payne JA (2004) Cation transport by the neuronal K(+)-Cl(-) cotransporter KCC2: thermodynamics and kinetics of alternate transport modes. Am J Physiol Cell Physiol 287:C919–C931

Wojcik SM, Katsurabayashi S, Guillemin I, Friauf E, Rosenmund C, Brose N, Rhee JS (2006) A shared vesicular carrier allows synaptic corelease of GABA and glycine. Neuron 50:575–587

Wong RK, Prince DA (1981) Afterpotential generation in hippocampal pyramidal cells. J Neurophysiol 45:86–97

Woo NS, Lu J, England R, McClellan R, Dufour S, Mount DB, Deutch AY, Lovinger DM, Delpire E (2002) Hyperexcitability and epilepsy associated with disruption of the mouse neuronal-specific K-Cl cotransporter gene. Hippocampus 12:258–268

Wu C, Asl MN, Gillis J, Skinner FK, Zhang L (2005) An in vitro model of hippocampal sharp waves: regional initiation and intracellular correlates. J Neurophysiol 94:741–753

Wu Y, Wang W, Diez-Sampedro A, Richerson GB (2007) Nonvesicular inhibitory neurotransmission via reversal of the GABA transporter GAT-1. Neuron 56:851–865

Yamada J, Okabe A, Toyoda H, Kilb W, Luhmann HJ, Fukuda A (2004) Cl- uptake promoting depolarizing GABA actions in immature rat neocortical neurones is mediated by NKCC1. J Physiol (Lond) 557:829–841

Yuste R, Katz LC (1991) Control of postsynaptic Ca2+ influx in developing neocortex by excitatory and inhibitory neurotransmitters. Neuron 6:333–344

Yuste R, Peinado A, Katz LC (1992) Neuronal domains in developing neocortex. Science 257:665–669
Zhang LL, Delpire E, Vardi N (2007) NKCC1 does not accumulate chloride in developing retinal neurons. J Neurophysiol 98:266–277
Zhou Q, Poo MM (2004) Reversal and consolidation of activity-induced synaptic modifications. Trends Neurosci 27:378–383
Zhu L, Lovinger D, Delpire E (2005) Cortical neurons lacking KCC2 expression show impaired regulation of intracellular chloride. J Neurophysiol 93:1557–1568

Chapter 8
Endocannabinoids and Inhibitory Synaptic Plasticity in Hippocampus and Cerebellum

Bradley E. Alger

8.1 Introduction

Endogenous cannabinoids (eCBs) are the natural ligands for the cannabinoid receptors in the brain. ECBs influence inhibitory synaptic plasticity through their receptors by shaping the formation of neuronal circuits, regulating the expression of synaptic plasticity, and influencing postsynaptic excitability. This chapter focuses on the cellular neurophysiology of eCB actions in inhibitory synaptic plasticity, although their effects at excitatory synapses will be touched upon.

8.1.1 Introduction to eCBs: History and Pharmacology

The story of the discoveries of eCBs, their receptors, and their functional roles, has been reviewed (see e.g., Howlett et al. 2002; Pertwee 2005; Di Marzo et al. 1998; Alger 2002; Freund et al. 2003; Piomelli 2003; Chevaleyre et al. 2006). Critical milestones included the isolation of delta-9 tetrahydrocannabinol (THC) as the major psychoactive component of the plant *Cannabis sativa* (Gaoni and Mechoulam 1964). Pharmacological agonist binding properties implied the existence of a specific cannabinoid receptor. The first cloned cannabinoid receptor, CB1R (Matsuda et al. 1990), proved to be the major CB receptor in the brain (CB2R is mainly present in certain glia (van Sickle et al. 2005) and is associated with the immune system). A new CB receptor, GPR55 (Ryberg et al. 2007), has been characterized pharmacologically, but not yet physiologically. CB1R is the most abundant heterotrimeric $G_{i/o}$-protein coupled receptor (GPCR) in the brain. The major intercellular

---

B.E. Alger (✉)
Departments of Physiology and Psychiatry, Program in Neuroscience, University of Maryland School of Medicine, 655 W Baltimore St, Baltimore, MD, 21201, USA
e-mail: balgerlab@gmail.com

endogenous ligands for CB1R are N-arachidonyl ethanolamine (anandamide or AEA; Devane et al. 1992) and 2-arachidonyl glycerol (2-AG; Mechoulam et al. 1995; Sugiura et al. 1995), although other candidates exist. It is not known why there are two endogenous ligands for CB1R. AEA has high affinity for CB1R but is a partial agonist; 2-AG is a full agonist with lower affinity. AEA is also a TRPV1 agonist that can directly affect 2-AG metabolism (Maccarrone et al. 2008). CB1R and its endogenous ligands constitute the cannabinoid system. Specific pharmacological tools, especially the receptor "antagonists" (actually inverse agonists) SR141617A (also called "rimonabant") and AM251, and agonists WIN55212-2 and CP55940, were vital to the discovery and elucidation of the numerous neurophysiological roles of the eCB system. Nevertheless, the one-time gold standard, rimonabant, is a TRPV1 antagonist as well (Gibson et al. 2008), and AM251 may be an agonist at GPR55 (Ryberg et al. 2007). Several lines of CB1R$^{-/-}$ mice have been developed, and suspected CB1R-mediated effects must be checked in a mutant mouse.

Compared with the elaborate vesicular secretory machinery for conventional neurotransmitters, the proposed signaling process employed by eCBs is simple. AEA and 2-AG are produced by enzymatic cleavage of lipid precursors in the outer membrane lipid bilayer of nerve cells. AEA is formed from N-arachidonyl phosphatidyl ethanolamine by the action of a phospholipase D, whereas 2-AG is derived from diacylgyercol by the action of diacylglycerol lipase (DGL). Controversy exists as to whether or not phospholipase C (PLC) activation is mandatory for the formation of 2-AG. PLC activity appears to be important when 2-AG levels are assayed neurochemically and in physiologically assays of eCBs produced via the activation of muscarinic cholinergic and metabotropic glutamatergic receptors (Hashimotodani et al. 2005). Neither AEA nor 2-AG are pre-packaged in membrane-bound structures, or stored in identified depots, and are said to be produced "on-demand." The analogy with modern manufacturing methods is widely used, but there are questions about how literally it should be interpreted. The process of eCB release is not understood.

AEA actions are terminated by the enzyme fatty-acid amide hydrolase (FAAH), and those of 2-AG are terminated by monoglyceride lipase (MGL). Both eCBs are taken up into cells by an as yet uncloned eCB-transporter. The uptake and degradation systems are very effective, and the natural agonists have only weak effects when bath-applied to brain slices. Synthetic agonists that are immune to transport and degradation are widely used instead.

The ubiquity of eCBs and their receptors, and the rapid ascent of the eCB system to a prominent place in modern neurophysiological research may obscure some fundamental, unresolved issues. A central problem is identifying the particular eCB that is active at a given synapse. Present technologies have limited spatial and temporal resolution. A general solution will probably have to await the development of new methods that permit real-time, in situ analyses of lipid signals.

8.2 Basic Neurophysiology of eCBs

8.2.1 *Retrograde Signaling*

The first suggestion that eCBs could be retrograde signals seems to have been based on theoretical rather than experimental grounds in prescient reports by Elphick and colleagues (Egertova et al. 1998; Elphick and Egertova 2001). Retrograde signals are produced in and released from a postsynaptic cell and then travel backwards across synaptic junctions where they activate receptors on presynaptic nerve terminals and alter synaptic transmitter release. Elphick and Ergetova noted that two major components of the eCB system, CB1Rs and the degradative enzyme for anandamide, FAAH, were expressed independently, with FAAH present in postsynaptic cells receiving inputs from pre-synaptic terminals bearing CB1R. They recognized that this organization was ideal for a retrograde signal system. Only the absence of an efficient means for testing this remarkable insight can explain the fact that it was not more widely recognized at first. In fact, the neurophysiological complement for their idea was being developed in parallel, but the physiological link was also unrecognized.

8.2.2 *Depolarization-Induced Suppression of Inhibition*

In the early 1990s an unusual mode of synaptic communication was discovered in in vitro cerebellar and hippocampal slices (Llano et al. 1991; Pitler and Alger 1992). It was found that increases in intracellular calcium ion concentration ($[Ca^{2+}]_i$) in the principal projection neurons cause a transient (tens of seconds at experimental temperatures) decrease in the amplitude of incoming, GABA-mediated inhibitory postsynaptic potentials or currents (IPSP/Cs). Repetitive action potential firing of the postsynaptic cell, or a brief postsynaptic depolarization, readily suppress the IPSP/Cs. This phenomenon became known as depolarization-induced suppression of inhibition, or DSI (Alger and Pitler 1995). DSI could not be accounted for by down-regulation or other modification of post-synaptic $GABA_A$ receptors. Instead, many studies suggested that the decreased IPSP/Cs reflect a decrease in GABA release, and that a presynaptic, pertussis toxin-sensitive G-protein coupled receptor (GPCR) is involved (Pitler and Alger 1994). The combination of postsynaptic induction and presynaptic expression imply that there must be a retrograde messenger from the principal cells to the interneurons.

The DSI messenger remained unknown until 2001 when several groups showed that it is an eCB (Wilson and Nicoll 2001; Ohno-Shosaku et al. 2001; Diana et al. 2002). Agonists of CB1R mimic and occlude DSI, and CB1R antagonists prevent it. A blocker of the eCB transporter suppresses IPSCs in a CB1R-dependent manner (Wilson and Nicoll 2001), suggesting that a low tonic level of CB1R activation is present. CB1R-dependent IPSC suppression is triggered by photo-uncaging of calcium in the postsynaptic cell. There is a tight correlation between the susceptibility of IPSC depression to DSI and suppression by the CB1R agonist WIN55212-2 in tissue cultured cells (Ohno-Shosaku et al. 2001). ECBs inhibit presynaptic release from GABAergic boutons measured by the FM1-43 destaining method (Brager et al. 2003). DSI is absent in two strains of CB1R$^{-/-}$ mice (Wilson et al. 2001; Varma et al. 2001), providing firm evidence for the involvement of eCBs in DSI.

The proposal that eCBs are the retrograde messengers in DSI fits well with the morphological localization of CB1Rs, which are expressed on the axon terminals of GABAergic interneurons in both cerebellum and hippocampus (Freund et al. 2003). In the hippocampus, the highest density of CB1Rs is on a subclass of GABAergic interneuron that also expresses the neuropeptide cholecystokinin (CCK) (Marsicano and Lutz 1999; Katona et al. 1999). The CCK cells comprise both basket cells and dendrite-targeting cells, and have distinctive properties, including an exclusive dependence on the conotoxin-sensitive, N-type voltage-gated calcium channels (VGCCs) for release of GABA. Coupled cell pair recordings revealed that conotoxin blocks all IPSCs that are susceptible to DSI or eCBs (Wilson et al. 2001). Presynaptic CB1Rs regulate release mainly by blocking VGCCs and reducing calcium influx into nerve terminals (Kreitzer and Regehr 2001b; Diana et al. 2002), although K channel activation and direct interference with the release processes can also contribute (Varma et al. 2002; Diana and Marty 2004). In cerebellum, inhibition of interneuron firing (Kreitzer et al. 2002), probably by increasing activity of a K channel, partly accounts for DSI.

ECB signaling is usually studied under non-physiological conditions, and it has been suggested that normal action potential firing patterns of CA1 cells are insufficient to cause eCB release (Hampson et al. 2003), although convergent, synchronous synaptic inputs from multiple sources are effective (Zhuang et al. 2005). Other physiologically relevant factors are considered below.

8.2.3 GPCR-Dependent eCB Mobilization

The evidence that DSI is initiated by a post-synaptic rise in $[Ca^{2+}]_i$ is persuasive: DSI is blocked by high concentrations of Ca^{2+} chelators, and is induced by stimulation that increases $[Ca^{2+}]_i$ in post-synaptic cells (Glitsch et al. 2000; Brenowitz and Regehr 2003; Wang and Zucker 2001), including photolytic uncaging of Ca^{2+} in these cells (Wilson and Nicoll 2001; Wang and Zucker 2001). NMDAR activation can also trigger Ca^{2+}-dependent eCB mobilization (Ohno-Shosaku et al. 2007). Early evidence had pointed to a close relationship between DSI and activation of certain G-protein coupled receptors (Pitler and Alger, 1994; Morishita et al. 1997; Martin and Alger 1999). After the discovery that DSI is mediated by eCBs, it was found that high concentrations of mGluR or mAChR agonists directly stimulate eCB mobilization (Maejima et al. 2001; Varma et al. 2001; Kim et al. 2002; Galante and Diana 2004). In addition, concentrations of GPCR agonists that are too low to stimulate detectable eCB effects directly, can enhance DSI (Varma et al. 2001, Kim et al. 2002; Brenowitz and Regehr 2005). The various forms of eCB mobilization are mediated by different intracellular biochemical cascades. Ca^{2+}-dependent eCB processes (DSI or depolarization-induced suppression of excitation (DSE), see below) are referred to as eCB_{Ca}, and mGluR- or mAChR-dependent forms are designated eCB_{mGluR} and eCB_{mAChR} (generally, eCB_{GPCR}).

It is not known if DSI enhancement by GPCRs is Ca^{2+}-dependent, because DSI itself requires Ca^{2+}. eCB_{GPCR} is relatively independent of $[Ca^{2+}]_i$ and can be initiated even when principal cells are loaded with Ca^{2+} chelators (Maejima et al. 2001; Kim et al. 2002). eCB_{GPCR} is not completely Ca^{2+} independent, however, and two models have been put forward to account for the role of Ca^{2+} in eCB_{GPCR}. The coincidence-detection model proposes that hippocampal $PLC_{\beta 1}$ (or $PLC_{\beta 4}$ in cerebellum, Maejima et al. 2005) integrates Ca^{2+} with a GPCR-induced intracellular messenger to generate eCBs (Hashimotodani et al. 2005). $PLC_\beta^{-/-}$ mice lose eCB_{GPCR} but not eCB_{Ca}. PLC_β isoforms are Ca^{2+}-dependent, and in this model $[Ca^{2+}]_i$ at levels $\geq 10^{-7}$ M is required to mobilize eCBs. A narrow window of time, set by the duration of $[Ca^{2+}]_i$ elevation, determines when eCBs are mobilized. The coincidence-detection model posits a single final common biochemical pathway for all eCB_{GPCR}. In the other model, the priming model, a transient rise in $[Ca^{2+}]_i$ is required to "prime" the initiation of eCB_{mGluR}, however, once the pathway is primed, a sustained elevation of $[Ca^{2+}]_i$ is not required, and eCB_{mGluR} can be generated at very low $[Ca^{2+}]_i$ (Edwards et al. 2008). In the priming model, PLC is upstream of the eCB_{GPCR} signaling step. Blockers of eCB mobilization, such as DGL inhibitors, prevent eCB_{mAChR} without reducing DSI enhancement by either mAChR or mGluR agonists, suggesting that the DSI enhancement and direct mobilization pathways do not share the same intracellular biochemical pathways. eCB_{mAChR} and eCB_{mGluR} can be distinguished in other ways as well (Edwards et al. 2006). In both models, Ca^{2+} and the products of GPCR activation interact non-linearly to mobilize eCBs (see also Brenowitz and Regehr 2005). Differences between mAChRs and mGluRs argue against the concept of a final common pathway for eCB mobilization (Edwards et al. 2008; see summary diagram in Fig. 8.1).

Regardless of the details, the synergistic interactions between $[Ca^{2+}]_i$ and GPCR products considerably broaden the scope and impact of eCBs in the brain. Although vigorous bursts of action potentials or depolarizations lasting hundreds of milliseconds and producing large increases in $[Ca^{2+}]_i$ undoubtedly occur, these are probably rare, whereas it is likely that moderate increases in $[Ca^{2+}]_i$ often overlap in time with GPCR activation. Hence, the integration of Ca^{2+} and GPCR products may be the most common mode of eCB mobilization.

8.2.4 Are eCBs Really Retrograde Messengers?

The answer to this question is almost certainly yes, nevertheless, the definitive evidence is not yet in. Unlike many conventional neurotransmitters which are

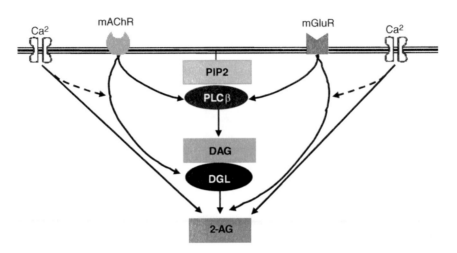

Fig. 8.1 Distinctions among DSI and eCB$_{mAChr}$ and eCB$_{mGluR}$ in hippocampus. *Arrows* denote known steps in signaling pathways for eCBs; unknown steps may be included within the *arrows*. DSI (eCB$_{Ca}$) and Ca^{2+} enhancement of eCB$_{GPCR}$ (*dotted arrows*) are independent of PLC. Intracellular injection of a PLC inhibitor does not affect any response. DSI is normal in PLC$_{\beta1}^{-/-}$ mice, but eCB$_{mAChR}$ and eCB$_{mGluR}$ are absent, suggesting PLC$_{\beta1}$ is upstream of eCB signaling. Intracellular DGL inhibitors have little effect on DSI or eCB$_{mGluR}$ but strongly reduce eCB$_{mAChR}$. Extracellular PLC or DGL inhibition prevents eCB-iLTD induction. Evidently different pathways are used for eCB mobilization under different conditions. Summary of data from several sources; see text

small, generally water-soluble molecules (glutamate, GABA, ACh, etc.), eCBs are hydrophobic lipids. They stick to membranes (and plastic tubing, glassware, and experimental chambers). They are effective in miniscule quantities, do not diffuse great distances in brain tissue (because of their hydrophobicity, as well as the aggressive uptake and degradation systems), and generally cannot be collected in the superfusate of stimulated physiological tissue. Direct methods of assaying eCBs rely on disruptive bulk treatment methods: e.g., lipid extraction followed by tandem mass spectroscopy. Relatively large quantities of brain tissue are required, and so temporal resolution and cellular specificity are lost. Furthermore, eCBs can be by-products of reactions that are unrelated to intercellular signaling. Indeed, a large proportion of the 2-AG detected by bulk assay methods probably serves functions other than signaling. Tissue stimulation increases the quantity of 2-AG that is detected neurochemically, but the signaling fraction cannot be separated from other fractions. The increases in 2-AG could reflect topping-up of the cellular reservoirs, and the on-demand production of eCBs might not be directly linked to signaling.

At the single cell level, the problems in assaying the eCB system are acute. Does cellular stimulation truly induce the de novo synthesis of eCBs, or does it only help to make eCBs available in some other way, e.g., by facilitating their exit from the cell of origin? Although appropriate stimulation of a postsynaptic cell clearly leads to a retrograde signaling process, which in turn leads to the activation of presynaptic CB1Rs, direct evidence of de novo postsynaptic synthesis of an eCB followed by its transit across the synapse is not yet available.

8.2.5 ECB Mobilization

The uncertainties alluded to above argue for caution. In particular, the implication that single cell stimulation "produces" or "releases" eCBs should probably be avoided until these processes can be directly measured. When stimulation of a postsynaptic cell initiates retrograde signaling that involves presynaptic CB1R receptors, the term "mobilization" can be used. This is descriptive of what is going on (the analogy is to antigenic mobilization of the immune response), but neutral as to the specific mechanisms (synthesis, release, transport, etc.) that are being triggered. As the details emerge, more specific terms will replace this general one.

8.2.6 Pre-endocannabinoid DSI and eCBs

Much was known about DSI before it was found to be mediated by eCBs (Alger 2002). Yet, to date, all of the discoveries made regarding DSI in the pre-eCB period have been verified by subsequent work. It is likely that all of the physiological evidence that DSI is expressed presynaptically applies also to demonstrated eCB-mediated processes, although not all of the experiments done on DSI have been repeated on the eCB actions. This chapter will not distinguish between pre-eCB DSI and eCB-mediated DSI when it is clear that pre-eCB phenomena translate readily to the eCB-mediated events.

8.2.7 Timing of eCB Mobilization

Timing of neuronal interactions is critical to the proper operation of the brain. Understanding the roles of any signaling system, including eCBs, requires understanding its temporal parameters. The physiological eCB action that is expressed as DSI does not peak until hundreds of milliseconds after a voltage step that causes Ca^{2+}-influx (Pitler and Alger 1994; Wilson and Nicoll; 2001). This is orders of magnitude longer than needed for conventional neurotransmission. DSI comprises all of the steps of eCB mobilization, and the activation of CB1Rs, which like all GPCRs, have comparatively slow actions. CB1Rs primarily target the downstream modulation of neurotransmitter release. To estimate the actual eCB mobilization time itself, Heinbockel et al. (2005) created a "caged" form of AEA. Caged AEA is biologically inactive until it is exposed to a brief flash of laser light that instantly (µs) disrupts the bond joining the chemical caging group with AEA, liberating bona fide AEA. Caged AEA can be applied to slices where, prior to laser flash, it equilibrates throughout the tissue immediately adjacent to CB1Rs. The diffusion time between photolytic AEA generation and CB1Rs is negligible, and therefore, the interval between the laser flash and the onset of IPSC suppression represents the time for binding of AEA to CB1R and downstream steps. This turns out to constitute a major fraction of the delay in IPSC depression that is seen in DSI or eCB_{GPCR}. The remaining steps normally leading to AEA synthesis and release must be relatively fast. It was concluded that eCBs are mobilized experimentally within 100 ms (at physiological temperatures ~50 ms) after the start of cellular stimulation. The brevity of this interval will influence the physiological roles played by eCBs, and constrain hypotheses about the underlying mobilization mechanisms.

8.2.8 2-AG is Probably the Main eCB in Hippocampus and Cerebellum

Determining whether or not eCBs are involved in a particular process is straightforward. Identifying which eCB is involved is more complicated. In the hippocampus, the first evidence that the signaling eCB was likely to be 2-AG came from chemical analyses of stimulated hippocampal slices showing selectively increased 2-AG levels (Stella et al. 1997). Kim and Alger (2004) found that in hippocampal slices an inhibitor of the AEA degradative enzyme, FAAH, did not affect DSI. Cyclooxygenase-2 (COX-2) (Kozak et al. 2000), a key inducible enzyme in prostaglandin synthesis, also degrades 2-AG. Inhibitors of COX-2 enhance DSI (Kim and Alger 2004; Sang et al. 2006; Hashimotodani et al. 2007) as expected if the enzyme helps to keep the 2-AG levels low. Increasing COX-2 levels decreases DSI (Sang et al. 2006). The effects of bath-applied 2-AG are not enhanced by COX-2 inhibition (Kim and Alger 2004; Hashimotodani et al. 2007), implying that COX-2 acts within the pyramidal cells, probably regulating the actual production of 2-AG, rather than serving as an elimination step for released 2-AG. The ineffectiveness of FAAH inhibition compared with the efficacy of the COX-2 inhibitors suggests that 2-AG mediates DSI.

Monoglyceride lipase (MGL) is a major degradative enzyme for 2-AG (Dinh et al. 2002), but not AEA, and its inhibition, therefore, should increase 2-AG mediated phenomena. MGL inhibitors enhance DSI (Makara et al. 2005; Hashimotodani et al. 2007). Bath-application of an inhibitor of DGL (the final enzyme in 2-AG synthesis), tetrahydrolipstatin, THL, also inhibits DSI (Hashimotodani et al. 2007), without affecting actions of applied 2-AG.

DGL is often localized at post-synaptic sites where eCBs may be liberated (Uchigashima et al. 2007; Yoshida et al. 2006). The main DGL isoform, DGL_α, is found in proximity to mGluRs in dendritic spines onto which CB1R-expressing excitatory terminals synapse in striatum (Uchigashima et al. 2007), cerebellum (Yoshida et al. 2006) and hippocampus (Yoshida et al. 2006; Katona et al. 2006).

PLC$_{\beta 4}$ (implicated in cerebellar eCB mobilization, Maejima et al. 2005) is also present in Purkinje cells (Yoshida et al. 2006). The pharmacological data, together with evidence of clustering of the major players in the eCB pathway, fulfill major criteria for identification of 2-AG as the eCB. DGL$_\alpha$ is apposed to inhibitory terminals in Purkinje cells, and in ventral tegmental area cells (Matyas et al. 2008). Caveats persist however. The arrangement just described does not always exist at inhibitory synapses. Yet, in the hippocampus and striatum, inhibitory synapses are much more sensitive to eCB actions than are the excitatory synapses. Activation of CB1Rs on glutamatergic terminals is proposed to be homosynaptic, whereas activation of CB1R on inhibitory terminals (i.e., by eCB$_{mGluR}$) would be heterosynaptic. Thus, the morphological arrangement raises key questions concerning eCB signaling at hippocampal inhibitory synapses, and suggests that different mechanisms underlie eCB mobilization in different parts of the brain (Fig. 8.2)

8.2.9 CB1R on Glutamatergic Terminals: Depolarization-Induced Suppression of Excitation

Although this chapter focuses on eCBs in inhibitory synaptic plasticity, eCBs also powerfully modulate glutamate transmission in many brain regions. Kreitzer and Regehr (2001b) reported that the retrograde suppression of glutamatergic synapses in the cerebellum, DSE, is mediated by an eCB. Both parallel and climbing fiber synapses are affected by DSE, and $[Ca^{2+}]_i$ -imaging experiments show that the

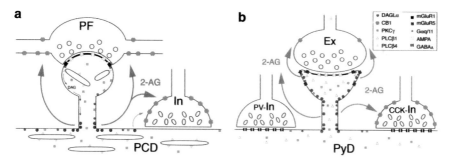

Fig. 8.2 Arrangements of the molecular components eCB$_{mGluR}$ in cerebellum (**a**) and hippocampal pyramidal cell dendrites (**b**) modified from Yoshida et al. (2006). In both regions interneuron terminals are more heavily invested with CB1Rs than excitatory terminals. Phospholipase Cs (β4 in cerebellum and β1 in hippocampus) are present in dendritic spines and shafts. In cerebellum, the 2-AG synthetic enzyme, DAGL$_\alpha$, is at the base of the spine and along the shaft. The metabotropic mGluR1 receptor is at the excitatory, PF, synapse, but is not near the inhibitory synapse. In the hippocampal pyramidal cell dendrite (PyD) DAGL$_\alpha$ is present throughout the spine, but not the dendritic shaft. mGluR5 is in hippocampus but not cerebellum, and both mGluR1 and mGluR5 are near the PyD excitatory synapses. The hippocampal CCK-In terminal bears CB1Rs, although the parvalbumin (PV-In) terminal does not. In both structures eCBs act homosynaptically on excitatory synapses and heterosynaptically on inhibitory synapses

eCBs depress glutamate release by depressing calcium influx into the synaptic terminals. DSE is induced by the same kinds of stimulation that induce DSI, and therefore, can participate in normal cerebellar network activity.

In the hippocampus, DSE is not produced by the same stimuli that induce DSI (Wagner and Alger 1996), except in autaptic culture (Straiker and Mackie 2005). Instead, much stronger postsynaptic stimulation (depolarization lasting ~10 s) is required to bring about a weak DSE (~15% EPSC reduction (Ohno-Shosaku et al. 2002; Chen et al. 2007). There is a much lower density of CB1Rs on excitatory terminals (Kawamura et al. 2006) than on inhibitory ones, although the greater number of excitatory terminals means that a significant fraction of the CB1Rs in the hippocampus could be on them. The diffuse but massive distribution of CB1Rs on glutamate terminals in hippocampus and neocortex may be primarily related to neuroprotection, as discussed below.

8.2.10 eCBs and Brain Development

8.2.10.1 eCBs Affect Interneuronal Connectivity

CB1Rs are heavily expressed in axonal growth cones of GABAergic interneurons in the rodent cortex during late gestation, where they help establish accurate connections between the interneurons and other cells. Interneuronal circuits are miswired in the CB1R$^{-/-}$ mouse (Berghuis et al. 2007), and AEA application inhibits interneuron neurite extension and opposes BDNF-induced neurite outgrowth (Berghuis et al. 2005). AEA also triggers CB1R internalization and elimination from filopodia of tissue-cultured GABAergic neurons (Berghuis et al. 2007). In tissue culture, AEA or WIN55212-2 induces chemorepulsion and collapse of the axonal growth cones of these GABAergic interneurons. Rho kinase inhibition prevents the effects caused by CB1R agonists. Interestingly, however, when a CB1R agonist is applied in the presence of the CB1R antagonist AM251, not only is axonal repulsion prevented, but the agonist becomes an attractant for axonal turning. Similarly, the Rho K antagonist converts AEA from chemo-repulsant to chemo-attractant. Neither effect has been explained. Nevertheless, eCBs can regulate synaptogenesis and target selection in vivo.

8.2.10.2 In Early Development eCBs Decrease Network Excitability

The high CB1R density on inhibitory nerve terminals, and the general rule that CB1R activation depresses transmitter release, strongly implies that eCBs will disinhibit network properties. (CB1R on excitatory terminals will have an opposite effect, but in hippocampus and neocortex these receptors are not easily activated by physiological stimuli.) However, at early postnatal developmental stages, up to about PN10 in rodents, postsynaptic GABA$_A$ receptors cause membrane depolarization (see Kaila et al. this volume), and have different roles than they do in adult tissue. Preventing the release of GABA would therefore decrease network excitability at these stages. Indeed, during early development, eCB-mediated retrograde signaling depresses network excitability by suppressing the excitatory GABA responses (Bernard et al. 2005). Conversely, CB1R antagonists cause epileptic discharges in the immature hippocampus. eCBs have, therefore, been proposed as mechanisms of homeostatic control of synaptic transmission in this tissue, capable of reducing or increasing network activity depending on the extent to which CB1Rs are activated. Since network activity is a crucial factor for the correct wiring of the brain, simple imbalances in the eCB system could adversely affect proper neuronal development.

These results raise concerns that activation of the eCB system by cannabis use during development could disturb synaptic development. Similar concerns would accompany the use of the CB1R antagonist (marketed as *Acomplia* in Europe) for weight loss. On the other hand, many centuries of experience with cannabis use by millions of people have evidently not lead to widespread major abnormalities attributable to brain mis-wiring (Iversen 2003). Cannabis consumption even during pregnancy was not associated with increased perinatal mortality or morbidity in a recent trial, although it was associated with a small, statistically detectable decrease in birthweight (Fergusson et al. 2002). The possibility of subtle effects cannot be dismissed, and such issues demand continued monitoring.

8.2.11 Interneurons Release eCBs

8.2.11.1 Interneuronal DSE and DSI

GABAergic interneurons also regulate their inputs by releasing eCBs. Cerebellar stellate and basket cells receive excitatory inputs from parallel fibers that express CB1Rs near their synaptic zones. Stimulation of the interneurons mobilizes eCBs and produces DSE (Beierlein and Regehr 2006). The phenomena in interneurons were essentially the same as in Purkinje cells, except that the stellate cell eCB-response is also triggered by NMDARs. Stellate cells inhibit Purkinje cells, hence stellate cell DSE influences feedforward inhibition onto the Purkinje cells. Decreasing the stellate cell inhibition in this way causes a much more widespread disinhibition than would be accomplished by DSI. In the hippocampus, the dendrite-targeting, Schaffer-Collateral Associated, CCK-expressing interneurons are electrically and chemically coupled. Their synapses express CB1Rs, and the SCA interneurons can induce DSI on each other's inputs (Ali 2007). Reducing inhibition of an inhibitory cell should result in a net stimulation of the principal cells, i.e., an influence opposite to that seen in cerebellum. Because a given interneuron typically activates hundreds of principal cells, eCB actions that alter interneuronal firing may be more globally dispersed than those that affect only principal cell firing.

8.2.11.2 eCB Mediated Self-Inhibition of Interneurons

Cortical low-threshold spiking (LTS) interneurons can release eCBs when vigorously stimulated (Bacci et al. 2004). Evidently, the eCBs act on CB1Rs on the LTS somata and increase a very long-lasting (>20 min) Ca^{2+}-dependent K^+-channel conductance, thus hyperpolarizing and inhibiting the cells. The implications of this effect for regulation of cortical networks are not understood, and the phenomenon has not been seen in other regions (e.g., Beierlein and Regehr 2006).

8.3 Basic Neurophysiology of eCBs and Synaptic Plasticity

8.3.1 Use-Dependent Regulation of eCB Effects on Inhibition

8.3.1.1 Increases in Probability of GABA Release Decrease Presynaptic eCB Effects

eCBs do not invariably and uniformly switch off GABA release. Their efficacy is a function of the activity in the interneuron, specifically in the synaptic terminal $[Ca^{2+}]_i$. Inhibition of voltage-gated N-type VGCCs, and the consequent decrease in $[Ca^{2+}]_i$ in the terminal, is the primary mechanism by which eCBs inhibit transmitter release in the short term. Increases in terminal $[Ca^{2+}]_i$, resulting from decreased K^+ conductance caused, e.g., by 4-aminopyridine (4-AP) (Alger et al. 1996; Morishita et al. 1998; Morishita and Alger 1999; Varma et al. 2002), can overcome DSI or eCB-induced IPSC suppression. 4-AP prolongs the terminal action potential, keeps presynaptic VGCCs open longer, and increases terminal $[Ca^{2+}]_i$. At low concentrations, $\leq 100\,\mu M$, 4-AP blocks only a few types of K^+ channels, which are often situated near nerve terminals. Extracellular application of other K^+ channel antagonists, i.e., TEA, Cs^+ or selective K^+-channel toxins, do not abolish DSI, suggesting that 4-AP sensitive channels have a privileged position in the CB1R-expressing inhibitory nerve terminals. As predicted, if Ca^{2+} influx via VGCCs is decreased, then 4-AP no longer abolishes DSI (Varma et al. 2002). Inhibition of GABA release caused by WIN55212-2 can also be overcome by 4-AP or barium in a $[Ca^{2+}]_o$-dependent way (Hoffman and Lupica 2000). Diana and Marty (2003) directly loaded the K^+ pore blocker Cs^+ into presynaptic interneurons, and observed a Ca^{2+}-sensitive reduction in cerebellar DSI. These data show that increasing the probability of GABA release can overcome the inhibitory effects of CB1R activation.

8.3.1.2 Tonic CB1R Activation

Physiological evidence of use-dependence of eCB effects comes from studies of inhibitory transmission between CCK-expressing mossy fiber associated interneurons and CA3 pyramidal cells (Losonczy and Nusser 2004). If stimulated at frequencies <25 Hz, the interneurons are essentially "mute," i.e., they produce almost no postsynaptic responses, but stimulation from 50 to 100 Hz elicits progressively more robust responses. Generally, the probability (Pr) of transmitter release is a function of presynaptic action potential firing frequency, and Pr increases with higher frequencies that cause higher $[Ca^{2+}]_i$. The muted interneuron synapses in CA3 have a low Pr at low stimulus frequencies. In this case, low Pr is not an intrinsic property of the synapses, rather it is caused by tonic activation of the CB1Rs on the presynaptic terminals. Inhibiting CB1Rs enables the cells to release GABA with a high Pr even at low stimulation frequencies. Either constitutive activation of the CB1R on the interneurons, or tonic release of eCBs in their vicinity, prevents them from releasing GABA. The output of nearby CCK-basket cells in CA3 is not similarly muted, implying that the eCB effects are somehow directed at the mossy fiber associated interneurons. In CA1, the output of CCK-basket cells in CA1 is also suppressed by tonic eCB actions (Neu et al. 2007). In CA1, increasing the presynaptic action potential firing to >20 Hz entirely reverses the DSI or inhibition of GABA release caused by a CB1R agonist. Apparently this frequency dependence, like the block of DSI produced by 4-AP, is attributable to increases in terminal $[Ca^{2+}]_i$ that accompany repetitive firing. Use-dependence of tonic eCB effects adds a new dimension to their ability to regulate the plasticity of inhibitory transmission.

8.3.1.3 Activity-Dependent Increases in eCB Responses

A different form of eCB use-dependence can be induced by low-frequency repetitive stimulation given for 5 min to afferent fibers in CA1 (Zhu and Lovinger 2007). Under this protocol, a slight degree of DSI is markedly and persistently enhanced by the activation of mGluRs during the stimulation. Edwards et al. (2008) found that a single DSI trial, causing a brief increase in pyramidal cell $[Ca^{2+}]_i$, enhances subsequent mGluR-mediated eCB-mobilization. Transient activation of mGluRs persistently enhances DSI, although prior activation of mAChRs does not. Therefore, the eCB system itself is plastic and subject to higher levels of regulation.

8.3.2 DSI in LTP

To appreciate the physiological roles of eCBs, it is necessary to consider their hydrophobic nature, and the powerful uptake and degradation systems that rapidly terminate their actions. Both factors retard their diffusion in the aqueous extracellular milieu, and severely limit the spread of eCBs from their source. Indeed, whether or not eCB released from one cell affects synaptic inputs to other cells near the source cell, is controversial (cf. Pitler and Alger 1994; Wilson and Nicoll 2001).

In general, eCBs act locally. This enables a single pyramidal cell to influence its own behavior without affecting other cells. Synaptic inhibition affects neuronal networks in many ways. For example, local disinhibition caused by DSI enables cells sourcing eCBs to opt out of communal activities and do things not being done by their neighbors. Reich et al. (2005) recorded from two pyramidal cells simultaneously, and observed that rhythmic IPSPs in one cell are not inhibited when the DSI is induced in the other cell (and vice versa) (Fig. 8.3a).

By hyperpolarizing cells, IPSPs help maintain the electrostatic Mg^{2+} ion plug of the NMDAR pore, keeping it in a non-conducting state. Disinhibition allows a given excitatory synaptic input to elicit greater than normal membrane depolarization, relief of the Mg^{2+} block, and expression of NMDAR-mediated responses. By restricting inhibition, eCBs should act as gating agents, and by acting locally in a time-limited way, contribute dimensions of temporal and spatial selectivity not provided by global modulation of GABA transmission. This model was tested by Carlson et al. (2002), who showed that DSI can facilitate LTP induction in CA1 pyramidal cells with simultaneous single whole-cell and field potential recordings. If timed to occur during a DSI period (disinhibition) in the single cell, a weak stimulus train of extracellular stimuli induces LTP only in the single cell, but not in the field potential. This shows that the disinhibition caused by DSI allows unblocking of NMDA receptors and LTP induction in a single cell, without affecting the population. In this way, DSI can target LTP to specific cells.

8.3.3 Inhibitory Long-Term Depression

Regulation of LTP by DSI (above) shows that eCBs can cause long-term effects indirectly, but does not rule out the possibility that eCBs might cause long-term synaptic modifications through a direct action. In the nucleus accumbens (Robbe et al. 2002) and striatum (Gerdeman et al. 2002), a form of long-term depression (LTD) is mediated by an eCB acting as a retrograde messenger at glutamatergic terminals. Both phenomena require activation of mGluRs and increases in $[Ca^{2+}]_i$ in the postsynaptic cells. An inhibitory LTD (iLTD) could be induced via a similar mechanism at GABAergic synapses in CA1 (Chevaleyre and Castillo 2003; Edwards et al. 2006). In the experiments, brief repetitive trains of afferent pathway stimulation release glutamate that activates mGluRs (ionotropic glutamate receptors were pharmacologically blocked; see Fig. 8.3b and c). Application of the group I mGluR-selective agonist DHPG for 10 min fully substitutes for afferent stimulation, and induces a chemical eCB-iLTD. iLTD induction is blocked by either mGluR or CB1R antagonists, but iLTD expression tested ~10 min post-induction is independent of both receptors. High concentrations of intracellular BAPTA in the pyramidal cells or extracellular (not intracellular, Edwards et al. 2006) application of either a PLC or a DGL inhibitor prevents iLTD induction. Importantly, LTP of EPSP-spike (E-S) coupling, i.e., of the ability of a given EPSP to trigger spikes, is

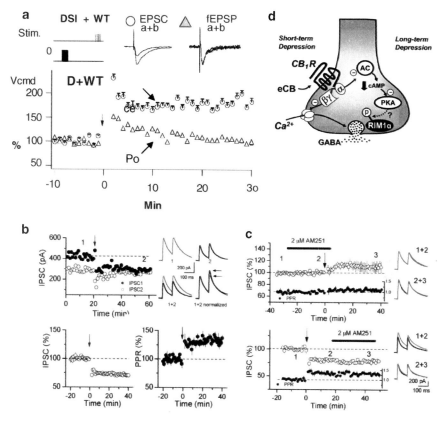

Fig. 8.3 eCBs depress inhibition and facilitate LTP induction. (**a**) A weak (0.4 s/50 Hz) stimulus train given to excitatory axons in s. radiatum of the hippocampal CA1 region does not induce LTP in the population field potential EPSP (fEPSP, *triangles* in graph). The EPSC recorded under voltage-clamp from a single cell in the same CA1 population undergoes LTP if the weak train is timed to coincide with the peak suppression of IPSCs produced by DSI in the cell (*open circles* in graph). Carlson et al. (2002). (**b**) (1) Initiation of iLTD in CA1 pyramidal cell by two strong 1-s/100 Hz stimulus trains in s. rad.; pairs of stimuli given througought. (**b**) (2) group data showing iLTD of first response of stimulus pair. (**b**) (3) Paired-pulse facilitation of the pair of IPSCs before and after iLTD induction. (**c**) A CB1R antagonist given before an iLTD protocol prevents iLTD induction (**c**) (1), but has no effect if given 10 min after induction (**c**) (2). (**b**) and (**c**) modified from Chevaleyre and Castillo (2003). (**d**) Diagram from Chevaleyre et al. (2007) illustrating the molecular sequelae of activating CB1R on the presynaptic interneuron terminal during DSI (**a**), and iLTD (**b** and **c**)

enhanced during iLTD (Chevaleyre and Castillo 2003). Thus, as with DSI-induced LTD, iLTD promotes LTP by suppressing IPSPs.

In order to investigate the spread of iLTD induction, Chevaleyre and Castillo (2004) focally stimulated in the CA1 dendritic region. Relatively weak stimulus trains delivered near a dendrite induced iLTD of DSI-sensitive IPSCs. If two stimulating electrodes were used, one to elicit the test IPSC, and the other to deliver the stimulus trains, the spread of the iLTD effect could be estimated by the distance along the dendrites between the two electrodes. iLTD induction affected inhibitory synapses only within ~10 μm of the stimulation site. By varying the stimulation protocol, the authors discovered conditions that produce iLTD, but not LTP of EPSCs. The data revealed that LTP of E-S coupling can be induced in the absence of EPSC LTP; i.e., iLTD is the most important factor in E-S coupling LTP (Fig. 8.4).

The iLTD induction process itself raises interesting questions. Bursts of synaptic stimulation lasting only seconds suffice for induction, but if bath-application of DHPG is used, it must be given for ~10 min before iLTD is established (Chevaleyre and Castillo 2003; Edwards et al. 2006). A similar prolonged activation of CB1R following synaptic stimulation probably takes place, because application of a CB1R antagonist beginning 3 min after stimulus trains lasting only seconds can completely prevent iLTD induction. Even 7 min afterwards, an antagonist partially reduces the iLTD magnitude (Chevaleyre and Castillo 2003; cf Ronesi et al. 2004), demonstrating that the CB1Rs must be activated for many minutes.

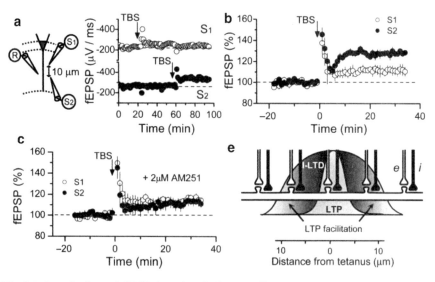

Fig. 8.4 Long-lasting and CB1R-dependent facilitatory effect on surrounding synapses. (**a**) Two stimulating electrodes were placed 10 μm apart in the middle third of s. radiatum along the apical dendrite of CA1 pyramidal cells and synaptic responses were recorded extracellularly. Stimulus strength was set to evoke identical synaptic responses in both pathways. Theta-burst stimulation (TBS) was first applied to one pathway (S1) and then to the other pathway (S2) 35 min later. (**b**) Group data ($n=5$) from experiments performed as described in (**a**). The first tetanus was delivered either to the proximal ($n=2$) or distal ($n=3$) stimulating pipettes. (**c**) Group data showing that LTP facilitation was abolished when TBS to S1 was delivered in the presence of 2 μM AM251 ($n=5$ slices). (**d**) Model that summarizes the local facilitatory effects of iLTD on LTP induction at Sch-CA1 synapses. Excitatory (e, *white*) and inhibitory (i, *black*) inputs impinge on the apical dendrite of a pyramidal cell. Local activation of excitatory inputs triggers LTP in a highly restricted area (10 μm from the stimulating site) and at the same time, it triggers iLTD in a slightly larger area. The spread of iLTD facilitates induction of LTP at neighboring excitatory inputs. Modified from Chevaleyre and Castillo (2004)

How does brief train stimulation give rise to prolonged eCB mobilization? In principle, a reduction in MGL activity could cause prolonged eCB effects (Hashimotodani et al. 2007). Alternatively perhaps, reversal of the eCB transporter (that normally mediates eCB uptake from the extracellular space) could result in secretion of eCBs into the extracellular space for an extended period of time. Ronesi et al. (2004) found that eCBs experimentally loaded into medium spiny neurons in the striatum are released in a transporter-dependent way. Indeed, the pattern of afferent stimulation determines the extent to which transporter-aided eCB secretion occurs (Adermark and Lovinger 2007). Double-pulse stimulation is much more effective than single-pulse stimulation in causing the transporter-dependent responses. How increasing the stimulation of the presynaptic cell would improve release of eCBs from the postsynaptic cell, or even if that happens, is unclear. Nevertheless, once transporter-dependent release is triggered, continued afferent stimulation is no longer required, suggesting that a long-lasting facilitation of eCB release is set into motion. Somewhat surprisingly, this protocol does not induce LTD or iLTD.

Is minutes-long stimulation of CB1R per se sufficient for iLTD induction? This is a controversial topic. Bath-application of WIN55212-2 alone reportedly induces iLTD, suggesting the CB1R activation alone is sufficient (Chevaleyre and Castillo 2003), but WIN55212-2 washes out only very slowly from tissue. On the other hand, the steady release of eCBs and minutes-long IPSC suppression resulting from injection of the G-protein activator $GTP_\gamma S$ into cells does not cause iLTD; as soon as AM251 is applied, the IPSCs return to control levels (Kim et al. 2002; Ronesi et al. 2004). Moreover, while an mGluR agonist induces eCB-iLTD, application of an mAChR agonist for ~20 min causes only a reversible, eCB-dependent IPSC suppression (Edwards et al. (2006). Finally, persistent activation of CB1Rs by a series of DSI trials that lasts for 10 min also fails to induce iLTD (Edwards et al. 2006). A similar controversy in the striatum regarding induction of eCB-LTD of glutamate synapses (cf. Ronesi et al. 2004 and Kreitzer and Malenka 2005), has been resolved by the report (Singla et al. 2007) that stimulation of the presynaptic neuron during the CB1R activation is mandatory for eCB-LTD induction. Hippocampal iLTD induction by repetitive DSI trials also requires simultaneous interneuron stimulation (Chevaleyre et al. 2007). Hence, a consensus seems to be emerging that activation of CB1R alone is insufficient to cause long-term synaptic plasticity, and that co-factors, perhaps in the nerve terminals, are required (cf. Edwards et al. 2006).

iLTD, but not DSI, requires presynaptic cAMP/PKA signaling, because it is inhibited by global inhibition of PKA, but not by injection of PKA inhibitors into the post-synaptic cells (Chevaleyre et al. 2007). iLTD, as well as the chemical eCB-iLTD induced by DHPG application, is absent in mutant mice lacking the active zone protein RIM1α. WIN55212-2 cannot induce iLTD in RIM1α$^{-/-}$ mice, however, RIM1α is not required for basal synaptic transmission or DSI. Similar results are obtained in amygdala and hippocampus. Although appealingly simple, the model summarized in Fig. 8.3d already requires updating. In a mouse in which RIM1α cannot be phosphorylated by PKA, iLTD is not blocked Kaeser et al (2008). Hence, the roles that RIM1α and PKA play in iLTD remain undetermined. The presynaptic activity of the serine/threonine phosphatase calcineurin (CaN) could be involved (Heifets et al. 2008). CaN activity is essential for iLTD expression, but whether or how it coordinates with the PKA/RIM1α scenario is also a mystery.

8.3.4 Relationship of the eCB System to Exogenous Cannabinoids

LTP is probably the neurophysiological underpinning for behavioral learning, and therefore agents that facilitate LTP should enhance learning. By inducing DSI and iLTD, eCBs facilitate LTP, yet cannabis use commonly impedes or disrupts learning, and exogenous cannabinoids can suppress LTP, e.g., (Sullivan 2000). Learning and LTP are highly complex phenomena, so resolution of these paradoxical findings will be multifaceted. Nevertheless, part of the explanation probably lies in the very different ways in which endogenous and exogenous cannabinoids affect the brain. The precisely localized, temporally and spatially constrained actions of the eCBs can be contrasted with cannabis use, in which CB1Rs are activated globally without regard to temporal or spatial limitations. Processes like LTP could even be facilitated at the cellular or synaptic level during cannabis use, but disruption of normal cellular and network patterning would alter normal storage and retrieval processes.

8.3.5 Spike-Timing Dependent Plasticity

The induction of many forms of synaptic plasticity depends on correlated spiking activity in presynaptic and postsynaptic cells. Whether spike-timing dependent plasticity (STDP) causes increases or decreases in synaptic strength depends critically on the temporal relationship between presynaptic and postsynaptic activation. Generally, if postsynaptic spikes repeatedly precede presynaptic transmitter release, then LTD is produced, and if transmitter release precedes the postsynaptic spikes, LTP occurs. Mobilization of eCBs is a major factor in STDP control at single dendritic spines from layer 2/3 pyramidal cells (Nevian and Sakmann 2006). A spike-timing dependent form of LTD (tLTD) at glutamate synapses in layer 5 pyramidal cells in the neocortex requires coincident postsynaptic and presynaptic activity as well as activation of presynaptic NMDARs (Sjostrom et al. 2003). Timing-dependent LTD induction depends on CB1R activation, but only within a narrow range of stimulation frequencies. The combined actions of presynaptic CB1Rs and NMDARs set the temporal window for tLTD induction.

eCBs also have a critical role in establishing STDP in cartwheel interneurons of the dorsal cochlear nucleus (Tzounopoulous et al. 2007). These cells follow an "antiHebbian" rule, whereby a presynaptic input that reliably induces spike firing induces LTD, rather than the expected LTP. The timing requirements for this LTD induction are extremely precise: an interval of only 5 ms between a presynaptic and a postsynaptic spike is required. The CB1R antagonist, AM251, prevents LTD induction and unmasks a conventional Hebbian form of LTP in the interneurons. Without the LTP process, LTD continues to be induced by EPSP-spike intervals of 20 ms. This suggests that the very narrow window for eCB-LTD depends on factors intrinsic to the LTD induction process, as well as on the increasing influence of an opposing LTP process. Predominance of either LTD or LTP is a function of afferent stimulation frequency, with lower frequencies favoring eCB-mediated LTD. The same afferent fibers making contacts onto fusiform cells in the nucleus are not subject to STDP LTD, apparently because CB1Rs are present in much lower density and in a different morphological arrangement at those sites. Evidently, mutual interactions between pre- and postsynaptic elements must occur during development in order to establish correct wiring of the eCB system.

Although eCBs regulate STDP, a puzzling and unresolved issue is how the actions of the eCBs are so tightly constrained in the temporal domain. The eCB system, while comparable in its speed of action to other GPCR-dependent signal systems, includes several steps, each of which is slower than the temporal requirements of STDP. It will be important to discover how this relatively slow system serves the much faster timing requirements of STDP.

8.3.6 eCBs and Seizures

Although the cerebellum does not undergo the abnormal hyperexcitability that characterizes epilepsy, the hippocampus is seizure-prone. In many epilepsy models, regulation of inhibition plays a key role in seizure initiation and propagation. Because they can inhibit GABA release, it might seem that cannabinoids would foster hyperexcitability, but generally this does this not happen. One reason is that not all interneurons express CB1Rs (Freund et al. 2003). CB1R-negative interneurons, a majority in most brain regions, continue to provide synaptic inhibition and thereby help prevent development of runaway excitability. Another major factor is that CB1Rs on excitatory nerve terminals suppress excitability by decreasing glutamate release (Marsicano et al. 2003).

Febrile seizures are fairly common in young children. Experimentally-induced febrile seizures persistently enhance DSI recorded in CA1 pyramidal cells (Chen et al. 2003), and seizures increase the tonic, i.e., unstimulated, activation of CB1Rs. Yet, the increase in eCB-mediated responses in post-seizure tissue is not attributable to increased eCB mobilization. Instead, febrile seizures up-regulate CB1Rs, as assessed by Western blots and a greater sensitivity to WIN55212-2. The number of CB1R-expressing nerve terminals does not increase, implying that CB1R density per terminal does. Tetanic stimulation mimicking seizure level activity in normal slices also up-regulates CB1Rs, via activation of AMPA/kainate receptors and mGluRs (Chen et al. 2007). The long-term increase in CB1R is prevented if a CB1R antagonist is present during the tetanus, implying that CB1Rs participate in their own up-regulation. The results have complex therapeutic implications: activation of CB1Rs, though usually anticonvulsant, could up-regulate CB1Rs on GABAergic terminals and have a pro-convulsant action in the long term because of eCB-silencing of the inhibitory synapses. Conversely, antagonism of CB1R during a seizure might cause a transient increase in excitability at that time, but prevent the long-term up regulation of CB1R, and thus be beneficial.

Under seizure conditions, vast numbers of principal neurons undergo strong stimulation and are at risk of excitotoxic damage in neocortex and hippocampus. The damage caused by kainic acid-induced seizures is intensified in $CB1R^{-/-}$ mice, implying that the eCB system is normally neuroprotective (Marsicano et al. 2003). eCBs that are profusely released during seizure activity have ready access to all CB1Rs. Do CB1Rs on GABAergic or on glutamatergic terminals mediate the neuroprotection? Studies on mutant mice with targeted deletions of CB1R on either GABAergic or glutamatergic neurons reveal that CB1Rs on glutamatergic neurons are fully responsible for eCB-mediated neuroprotection (Monory et al. 2006). Spread of excitotoxic damage is as extensive if the CB1R deletion is confined to the glutamatergic cells as it is in the global $CB1R^{-/-}$ animals. Selective deletion of CB1R from the GABAergic cells does not alter neuronal damage. Restriction of CB1R deletion to hippocampal dentate gyrus by injecting CRE-expressing virus into this region in CB1R-floxed mice leads to the same conclusion (Monory et al. 2005). Evidently CB1R activation on glutamatergic terminals limits further release of glutamate and, thereby, limits the extent of cell loss.

Status epilepticus (SE) is the extreme form of epileptiform hyperexcitability. Whereas normal seizures last from seconds to minutes, SE is a state of seizure activity that can last for ≥30 min, a major medical emergency that can lead to death. In a low-Mg^{2+} seizure model in hippocampal culture, CB1R receptor antagonists cause the development of continuous epileptiform activity that resembles SE (Deshpande et al. 2007a). The SE-like activity can be overcome by high concentrations of CB1R agonists. Control neurons treated with CB1R receptor antagonists do not undergo SE or hyperexcitability. Moreover, application of CB1R agonists can stop experimental SE in the same tissue culture model (Deshpande et al. 2007b). These findings suggest that endogenous eCBs can modulate seizure frequency and duration, and prevent the development of SE-like activity in epileptic neurons.

8.4 Development and eCBs

Does the eCB system remain the same across the developmental spectrum? The reduction in parallel and climbing fiber synaptic transmission caused by Purkinje cell activation is evidently exclusively mediated by an eCB (Kreitzer and Regehr 2001a). Yet, initial reports suggested that the primary retrograde messenger at the cerebellar Purkinje cell-parallel fiber synapses is glutamate, released from the Purkinje cell dendrites (Levenes et al. 2001). The discrepancy could reflect a developmental shift: in young animals the retrograde EPSC suppression could be entirely mediated by eCBs, but in older animals a mix of CB1R and mGluR could mediate retrograde signaling (Crepel 2007). There does not appear to be comparable information on a similar shift in the regulation of cerebellar GABAergic synapses, or on the regulation of synapses in other brain regions.

8.5 Synergy with Nitric Oxide System

Nitric oxide (NO) is a gaseous molecule produced by the Ca^{2+}-dependent activation of nitric oxide synthase (NOS). In hippocampus and cerebellum the possibility of synergistic interactions between eCB- and NO-mediated signaling exists, although the particulars differ. LTD of excitatory synapses in the cerebellar cortex is a postsynaptic phenomenon, mediated by down-regulation of AMPA receptors at the parallel fiber-Purkinje cell synapse. NO appears to be a key component of the LTD mechanism (e.g., Lev-Ram et al. 1997). Presynaptic CB1Rs on the excitatory synapses suppress glutamate release when activated by eCBs from the Purkinje cells. eCBs mediate LTD induction at parallel fiber synapses, but this is prevented by the NOS inhibitor, L-NAME (Safo and Regehr 2005), suggesting that NO is downstream of CB1R activation in cerebellum.

In hippocampus NO has been put forward as a parallel retrograde signaling messenger between the CA1 pyramidal cells and interneurons (Makara et al. 2007). NO produced in the pyramidal cells reportedly inhibits GABA release during mAChR activation. Neuronal nNOS is found in pyramidal cells at sites opposite to GABAergic synapses (Szabadits et al. 2007). The molecular receptor for NO, nitric oxide (soluble) guanylate cylase (NOsGC), is localized to nNOS-expressing GABAergic nerve terminals. Moreover, the α1 isoform of NOsGC is found exclusively in interneurons. Inhibition of either NO or eCB signaling almost entirely abolishes DSI (Makara et al. 2007). nNOS inhibitors, or scavenging NO with chelators such as CPTIO, significantly reduce DSI recorded in the presence of mAChR agonists. Although the data are intriguing, questions remain. A close association of α1β1 subunits of NOsGC with interneuron terminals is not obviously consistent with the eCB mechanism. Whereas most CCK- and PV-positive interneurons are NOsGC α1 positive, the CB1R receptor is uniquely localized on CCK interneurons, and specifically excludes the PV cells. NO has not been shown to affect the PV cells, hence its presence there is enigmatic.

Both NO and eCBs could affect DSI by acting in parallel or in series. They could target the same cells and their effects would summate. Alternatively, NO and eCBs might interact non-linearly; one could be upstream of the other, and their pathways could merge. In CA1, the latter situation appears to hold: blocking NOS, for example, almost entirely abolishes DSI (Makara et al. 2007). Yet, DSI is absent in CB1R$^{-/-}$ mice (Varma et al. 2001; Wilson et al 2001), implying that CB1R is the final common pathway for DSI, and that the NO and eCB pathways merge. But NO is not required for CB1R activation and has not been reported to stimulate eCB mobilization. Finally, in the absence of mAChR activation, DSI is not affected by the NO pathway (Makara et al. 2007). Apparently, mAChRs bring about a switch from a DSI mechanism that is NO-independent and CB1R-dependent to one in which NO and eCBs act interdependently. Despite complexities, the prospect of NO-CB1R interactions is interesting and will stimulate further investigation.

8.6 Conclusions

Inhibition shapes and regulates neuronal activity, and plasticity of inhibitory synapses is, therefore, an issue of broad significance. eCBs are important intercellular signaling molecules that operate differently in different brain areas. There is diffuse but extensive expression of CB1Rs on excitatory terminals in hippocampus, however, the major physiological targets of eCBs in hippocampus are the inhibitory interneurons that express the highest densities of CB1R. In hippocampus CB1Rs on excitatory terminals serve mainly as a fail-safe backup system, suppressing hyperexcitability during abnormal activity that liberates large quantities of glutamate. In cerebellum and other regions, physiologically released eCBs powerfully regulate both excitatory and inhibitory systems. mGluR-dependent eCB mobilization can induce LTD at many synapses. Short and long-term forms of eCB-dependent synaptic plasticity are ubiquitous. The existence of eCB systems throughout the brain indicates that there is much to be learned about how eCBs regulate inhibitory synaptic plasticity.

Acknowledgments Work in the author's laboratory is supported by NIH RO1 DA01465 and MH077277.

References

Adermark L, Lovinger DM (2007) Retrograde endocannabinoid signaling at striatal synapses requires a regulated postsynaptic release step. Proc Natl Acad Sci USA 104:20564–20569

Alger BE (2002) Retrograde signaling in the regulation of synaptic transmission: focus on endocannabinoids. Prog Neurobiol 68:247–286

Alger BE, Pitler TA (1995) Retrograde signaling at GABAA-receptor synapses in the mammalian CNS. Trends Neurosci 18:333–340

Alger BE, Pitler TA, Wagner JJ, Martin LA, Morishita W, Kirov SA, Lenz RA (1996) Retrograde signalling in depolarization-induced suppression of inhibition in rat hippocampal CA1 cells. J Physiol (Lond) 496:197–209

Ali AB (2007) Presynaptic inhibition of GABAA receptor mediated unitary IPSPs by cannabinoid receptors at synapses between CCK-positive interneurons in rat hippocampus. J Neurophysiol 98:861–869

Bacci A, Prince DA, Huguenard JR (2004) Long-lasting self-inhibition of neocortical interneurons mediated by endocannabinoids. Nature 431:312–316

Beierlein M, Regehr WG (2006) Local interneurons regulate synaptic strength by retrograde release of endocannabinoids. J Neurosci 26:9935–9943

Berghuis P, Dobszay MB, Wang X, Spano S, Ledda F, Sousa KM, Schulte G, Ernfors P, Mackie K, Paratcha G, Hurd YL, Harkany T (2005) Endocannabinoids regulate interneuron migration and morphogenesis by transactivating the TrkB receptor. Proc Natl Acad Sci USA 102:19115–19120

Berghuis P, Rajnicek AM, Morozov YM, Ross RA, Mulder J, Urban GM, Monory K, Marsicano G, Matteoli M, Canty A, Irving AJ, Katona I, Yanagawa Y, Rakic P, Lutz B, Mackie K, Harkany T (2007) Hardwiring the brain: endocannabinoids shape neuronal connectivity. Science 316:1212–1216

Bernard C, Milh M, Morozov YM, Ben-Ari Y, Freund TF, Gozlan H (2005) Altering cannabinoid signaling during development disrupts neuronal activity. Proc Natl Acad Sci USA 102:9388–9393

Brager DH, Luther PW, Erdélyi F, Szabó G, Alger BE (2003) Regulation of exocytosis from single visualized GABAergic boutons in hippocampal slices. J Neurosci 23:10475–10486

Brenowitz SD, Regehr WG (2003) Calcium dependence of retrograde inhibition by endocannabinoids at synapses onto Purkinje cells. J Neurosci 23:6373–6384

Brenowitz SD, Regehr WG (2005) Associative short-term synaptic plasticity mediated by endocannabinoids. Neuron 45:419–431

Carlson GC, Wang Y, Alger BE (2002) Endocannabinoids facilitate the induction of LTP in the hippocampus. Nat Neurosci 5:723–724

Chen K, Ratzliff A, Hilgenberg L, Gulyas A, Freund TF, Smith M, Dinh TP, Piomelli D, Mackie K, Soltesz I (2003) Long-term plasticity of endocannabinoid signaling induced by developmental febrile seizures. Neuron 39:599–611

Chen K, Neu A, Howard AL, Foldy C, Echegoyen J, Hilgenberg L, Smith M, Mackie K, Soltesz I (2007) Prevention of plasticity of endocannabinoid signalling inhibits persistent limbic hyperexcitability caused by developmental seizures. J Neurosci 27:46–58

Chevaleyre V, Castillo PE (2003) Heterosynaptic LTD of hippocampal GABAergic synapses. A novel role of endocannabinoids in regulating excitability. Neuron 38:461–472

Chevaleyre V, Castillo PE (2004) Endocannabinoid-mediated metaplasticity in the hippocampus. Neuron 43:871–881

Chevaleyre V, Takahashi KA, Castillo PE (2006) Endocannabinoid-mediated synaptic plasticity in the CNS. Annu Rev Neurosci 29:37–76

Chevaleyre V, Heifets BD, Kaeser PS, Sudhof TC, Purpura DP (2007) Endocannabinoid-mediated long-term plasticity requires cAMP/PKA signaling and RIM1α. Neuron 54:801–812

Crepel F (2007) Developmental changes in retrograde messengers involved in depolarization-induced suppression of excitation at parallel fiber-Purkinje cell synapses in rodents. J Neurophysiol 97:824–836

Deshpande LS, Sombati S, Blair RE, Carter DS, Martin BR, DeLorenzo RJ (2007a) Cannabinoid CB1 receptor antagonists cause status epilepticus-like activity in the hippocampal neuronal culture model of acquired epilepsy. Neurosci Lett 411:11–16

Deshpande LS, Blair RE, Ziobro JM, Sombati S, Martin BR, DeLorenzo RJ (2007b) Endocannabinoids block status epilepticus in cultured hippocampal neurons. Eur J Pharmacol 558:52–59

Devane WA, Hanus L, Breuer A, Pertwee RG, Stevenson LA, Griffin G, Gibson D, Mandelbaum A, Etinger A, Mechoulam R (1992) Isolation and structure of a brain constituent that binds to the cannabinoid receptor. Science 258:1946–1949

Di Marzo V, Melck D, Bisogno T, De Petrocellis L (1998) Endocannabinoids: endogenous cannabinoid receptor ligands with neuromodulatory action. Trends Neurosci 21:521–528

Diana MA, Marty A (2003) Characterization of depolarization-induced suppression of inhibition using paired interneuron – Purkinje cell recordings. J Neurosci 23:5906–5918

Diana MA, Marty A (2004) Endocannabinoid-mediated short-term synaptic plasticity: depolarization-induced suppression of inhibition (DSI) and depolarization-induced suppression of excitation (DSE). Br J Pharmacol 142:9–19

Diana MA, Levenes C, Mackie K, Marty A (2002) Short-term retrograde inhibition of GABAergic synaptic currents in rat Purkinje cells is mediated by endogenous cannabinoids. J Neurosci 22:200–208

Dinh TP, Carpenter D, Leslie FM, Freund TF, Katona I, Sensi SL, Kathuria S, Piomelli D (2002) Brain monoglyceride lipase participating in endocannabinoid inactivation. Proc Natl Acad Sci USA 99:10819–10824

Edwards DA, Kim J, Alger BE (2006) Multiple mechanisms of endocannabinoid response initiation in hippocampus. J Neurophysiol 95:67–75

Edwards DA, Zhang L, Alger BE (2008) Metaplastic control of the endocannabinoid system at inhibitory synapses in hippocampus. Proc Natl Acad Sci USA 105:8142–8147

Egertova M, Giang DK, Cravatt BF, Elphick MR (1998) A new perspective on cannabinoid signalling: complementary localization of fatty acid amide hydrolase and the CB1 receptor in rat brain. Proc R Soc Lond B Biol Sci 265:2081–2085

Elphick MR, Egertova M (2001) The neurobiology and evolution of cannabinoid signalling. Phil Trans R Soc Lond B 356:381–408

Fergusson DM, Horwood LJ, Northstone K, ALSPAC Study Team. Avon Longitudinal Study of Pregnancy and Childhood (2002) Maternal use of cannabis and pregnancy outcome. BJOG 109(1):21–27

Foldy C, Neu A, Jones MV, Soltesz I (2006) Presynaptic, activity-dependent modulation of cannabinoid type 1 receptor-mediated inhibition of GABA release. J Neurosci 26:1465–1469

Freund TF, Katona I, Piomelli D (2003) Role of endogenous cannabinoids in synaptic signaling. Physiol Rev 83:1017–1066

Galante M, Diana MA (2004) Group I metabotropic glutamate receptors inhibit GABA release at interneuron-Purkinje cell synapses through endocannabinoid production. J Neurosci 24:4865–4874

Gaoni Y, Mechoulam R (1964) Isolation, structure and partial synthesis of an active constituent of hashish. J Am Chem Soc 86:1646–1647

Gerdeman GL, Ronesi J, Lovinger DM (2002) Postsynaptic endocannabinoid release is critical to long-term depression in the striatum. Nat Neurosci 5:446–451

Gibson HE, Edwards JG, Page RS, Van Hook MJ, Kauer JA (2008) TRPV1 channels mediate long-term depression at synapses on hippocampal interneurons. Neuron 57:746–759

Glitsch M, Parra P, Llano I (2000) The retrograde inhibition of IPSCs in rat cerebellar Purkinje cells is highly sensitive to intracellular Ca2+. Eur J Neurosci 12:987–993

Hampson RE, Zhuang SY, Weiner JL, Deadwyler SA (2003) Functional significance of cannabinoid-mediated, depolarization induced suppression of inhibition (DSI) in the hippocampus. J Neurophysiol 90:55–64

Hashimotodani Y, Ohno-Shosaku T, Tsubokawa H, Ogata H, Emoto K, Maejima T, Araishi K, Shin H-S, Kano M (2005) Phospholipase Cb serves as a coincidence detector through its Ca2+ dependency for triggering retrograde endocannabinoid signal. Neuron 45:257–268

Hashimotodani Y, Ohno-Shosaku T, Kano M (2007) Presynaptic monoacylglycerol lipase activity determines basal endocannabinoid tone and terminates retrograde endocannabinoid signaling in the hippocampus. J Neurosci 27:1211–1219

Heifets BD, Chevaleyre V, Castillo PE (2008) Interneuron activity controls endocannabinoid-mediated presynaptic plasticity through calcineurin. Proc Natl Acad Sci USA 105:10250–10255

Heinbockel T, Brager DH, Reich CG, Zhao J, Muralidharan S, Alger BE, Kao JPY (2005) Endocannabinoid signaling dynamics probed with optical tools. J Neurosci 25:9449–9459

Hoffman AF, Lupica CR (2000) Mechanisms of cannabinoid inhibition of GABAA synaptic transmission in the hippocampus. J Neurosci 20:2470–2479

Howlett AC, Barth F, Bonner TI, Cabral G, Casellas P, Devane WA, Felder CC, Herkenham M, Mackie K, Martin BR, Mechoulam R, Pertwee RG (2002) International Union of Pharmacology. XXVII. Classification of cannabinoid receptors. Pharmacol Rev 54:161–202

Iversen L (2003) Cannabis and the brain. Brain 126:1252–1270

Kaeser PS, Kwon HB, Blundell J, Chevaleyre V, Morishita W, Malenka RC, Powell CM, Castillo PE, Südhof TC (2008) RIM1alpha phosphorylation at serine-413 by protein kinase A is not required for presynaptic long-term plasticity or learning. Proc Natl Acad Sci USA 105:14680–14685

Katona I, Sperlagh B, Sik A, Kafalvi A, Vizi ES, Mackie K, Freund TF (1999) Presynaptically located CB1 cannabinoid receptors regulate GABA release from axon terminals of specific hippocampal interneurons. J Neurosci 19:4544–4558

Katona I, Urban GM, Wallace M, Ledent C, Jung K-M, Piomelli D, Mackie K, Freund TF (2006) Molecular composition of the endocannabinoid system at glutamatergic synapses. J Neurosci 26:5628–5637

Kawamura Y, Fukaya M, Maejima T, Yoshida T, Miura E, Watanabe M, Ohno-Shosaku T, Kano M (2006) The CB1 cannabinoid receptor is the major cannabinoid receptor at excitatory presynaptic sites in the hippocampus and cerebellum. J Neurosci 26:2991–3001

Kim J, Isokawa M, Ledent C, Alger BE (2002) Activation of muscarinic acetylcholine receptors enhances the release of endogenous cannabinoids in the hippocampus. J Neurosci 22:10182–10191

Kim J, Alger BE (2004) Inhibition of cyclooxygenase-2 potentiates retrograde endocannabinoid effects in hippocampus. Nat Neurosci 7:697–698

Kozak KR, Rowlinson SW, Marnett LJ (2000) Oxygenation of the endocannabinoid, 2-arachidonylglycerol, to glyceryl prostaglandins by cyclooxygenase-2. J Biol Chem 275:33744–33749

Kreitzer AC, Malenka RC (2005) Dopamine modulation of state-dependent endocannabinoid release and long-term depression in the striatum. J Neurosci 25:10537–10545

Kreitzer AC, Regehr WG (2001a) Retrograde inhibition of presynaptic calcium influx by endogenous cannabinoids at excitatory synapses onto Purkinje cells. Neuron 29:717–727

Kreitzer AC, Regehr WG (2001b) Cerebellar depolarization-induced suppression of inhibition is mediated by endogenous cannabinoids. J Neurosci 21:RC174(1–5)

Kreitzer AC, Carter AG, Regehr WG (2002) Inhibition of interneuron firing extends the spread of endocannabinoid signaling in the cerebellum. Neuron 34:787–796

Levenes C, Daniel H, Crepel F (2001) Retrograde modulation of transmitter release by postsynaptic subtype 1 metabotropic glutamate receptors in the rat cerebellum. J Physiol (Lond) 537:125–140

Lev-Ram V, Jiang T, Wood J, Lawrence DS, Tsien RY (1997) Synergies and coincidence requirements between NO, cGMP, and Ca^{2+} in the induction of cerebellar long-term depression. Neuron 18:1025–1038

Llano I, Leresche N, Marty A (1991) Calcium entry increases the sensitivity of cerebellar Purkinje cells to applied GABA and decreases inhibitory synaptic currents. Neuron 6:565–574

Losonczy A, Nusser Z (2004) Persistently active cannabinoid receptors mute a sub-population of hippocampal interneurons. Proc Natl Acad Sci USA 101:1362–1367

Maccarone M, Rossi S, Bari M, De Chiara V, Fezza F, Musella A, Gasperi V, Prosperetti C, Bernardi G, Finazzi-Agrò A, Cravatt BF, Centonze D (2008) Anandamide inhibits metabolism and physiological actions of 2-arachidonoylglycerol in the striatum. Nat Neurosci 11:152–159

Maejima T, Hasimoto K, Yoshida T, Aiba A, Kano M (2001) Presynaptic inhibition caused by retrograde signal from metabotropic glutamate to cannabinoid receptors. Neuron 31:463–475

Maejima T, Oka S, Hashimotodani Y, Ohno-Shosaku T, Aiba A, Wu D, Waku K, Sugiura T, Kano M (2005) Synaptically driven endocannabinoid release requires Ca2+-assisted metabotropic glutamate receptor subtype 1 to phospholipase C β4 signaling cascade in the cerebellum. J Neurosci 25:6826–6835

Makara JK, Mor M, Fegley D, Szabo SI, Kathuria S, Astarita G, Duranti A, Tontini A, Tarzia G, Rivara S, Freund TF, Piomelli D (2005) Selective inhibition of 2-AG hydrolysis enhances endocannabinoid signaling in hippocampus. Nat Neurosci 8:1139–1141

Makara JK, Katona I, Nyíri G, Németh B, Ledent C, Watanabe M, de Vente J, Freund TF, Hájos N (2007) Involvement of nitric oxide in depolarization-induced suppression of inhibition in hippocampal pyramidal cells during activation of cholinergic receptors. J Neurosci 27:10211–10222

Marsicano G, Lutz B (1999) Expression of the cannabinoid receptor CB1 in distinct neuronal subpopulations in the adult mouse forebrain. Eur J Neurosci 11:4213–4225

Marsicano G, Goodenough S, Monory K, Hermann H, Eder M, Cannich A, Azad SC, Cascio MG, Gutiérrez SO, van der Stelt M, López-Rodriguez ML, Casanova E, Schütz G, Zieglgänsberger W, Di Marzo V, Behl C, Lutz B (2003) CB1 cannabinoid receptors and on-demand defense against excitotoxicity. Science 302(5642):84–88

Martin LA, Alger BE (1999) Muscarinic facilitation of the occurrence of depolarization-induced suppression of inhibition in rat hippocampus. Neuroscience 92:61–71

Matsuda LA, Lolait SJ, Brownstein MJ, Young AC, Bonner TI (1990) Structure of a cannabinoid receptor and functional expression of the cloned cDNA. Nature 346:561–564

Mátyás F, Urbán GM, Watanabe M, Mackie K, Zimmer A, Freund TF, Katona I (2008) Identification of the sites of 2-arachidonoylglycerol synthesis and action imply retrograde endocannabinoid signaling at both GABAergic and glutamatergic synapses in the ventral tegmental area. Neuropharmacol 54:95–107

Mechoulam R, Ben-Shabat S, Hanus L, Ligumsky M, Kaminski NE, Schatz AR, Gopher A, Almog S, Martin BR, Compton DR, Pertwee RG, Griffin G, Bayewitch M, Barg J, Vogel Z (1995) Identification of an endogenous 2-monoglyceride, present in canine gut, that binds to cannabinoid receptors. Biochem Pharmacol 50:83–90

Monory K, Massa F, Egertova M, Eder M, Blaudzun H, Westenbroek R, Kelsch W, Jacob W, Marsch R, Ekker M, Long J, Rubenstein J, Goebbels S, Nave KA, During M, Klugmann M, Wolfel B, Dodt H-U, Zieglgansberger W, Wotjak CT, Mackie K, Elphick MR, Marsicano G, Lutz B (2006) The endocannabinoid system controls key epileptogenic circuits in the hippocampus. Neuron 51:455–466

Morishita W, Alger BE (1999) Evidence for endogenous excitatory amino acids as mediators in DSI of GABAAergic transmission in hippocampal CA1. J Neurophysiol 82:2556–2564

Morishita W, Kirov SA, Pitler TA, Martin LA, Lenz RA, Alger BE (1997) N-Ethylmaleimide blocks depolarization-induced suppression of inhibition and enhances GABA release in the rat hippocampal slice in vitro. J Neurosci 17:941–950

Morishita W, Kirov SA, Alger BE (1998) Evidence for metabotropic glutamate receptor activation in the induction of depolarization-induced suppression of inhibition in hippocampal CA1. J Neurosci 18:4870–4882

Neu A, Foldy C, Soltesz I (2007) Postsynaptic origin of CB1-dependent tonic inhibition of GABA release at CCK-positive basket cell to pyramidal cell synapses in the CA1 region of the rat hippocampus. J Physiol (Lond) 578:233–247

Nevian T, Sakmann B (2006) Spine Ca2+ signalling in spike-timing-dependent plasticity. J Neurosci 26:11001–11013

Ohno-Shosaku T, Maejima T, Kano M (2001) Endogenous cannabinoids mediate retrograde signals from depolarized postsynaptic neurons to presynaptic terminals. Neuron 29:729–738

Ohno-Shosaku T, Tsubokawa H, Mizushima I, Yoneda N, Zimmer A, Kano M (2002) Presynaptic cannabinoid sensitivity is a major determinant of depolarization-induced retrograde suppression at hippocampal synapses. J Neurosci 22:3864–3872

Ohno-Shosaku T, Hashimotodani Y, Ano M, Takeda S, Tsubokawa H, Kano M (2007) Endocannabinoid signaling triggered by NMDA receptor-mediated calcium entry into rat hippocampal neurons. J Physiol (Lond) 584:407–418

Pertwee RG (2005) Pharmacology of cannabinoid CB1 and CB2 receptors. Pharmacol Ther 74:129–180

Piomelli D (2003) The molecular logic of endocannabinoid signaling. Nat Rev Neurosci 4:873–884

Piomelli D, Giuffrida A, Calignano A, Rodriguez de Fonseca F (2000) The endocannabinoid system as a target for therapeutic drugs. Trends Pharmacol Sci 21:218–224

Pitler TA, Alger BE (1992) Postsynaptic spike firing reduces synaptic GABAA responses in hippocampal pyramidal cells. J Neurosci 12:4122–4132

Pitler TA, Alger BE (1994) Depolarization-induced suppression of GABAergic inhibition in rat hippocampal pyramidal cells: G protein involvement in a presynaptic mechanism. Neuron 13:1447–1455

Reich CG, Karson MA, Karnup SV, Jones L, Alger BE (2005) Regulation of IPSP theta rhythms by muscarinic receptors and endocannabinoids in hippocampus. J Neurophysiol 94:4290–4299

Robbe D, Kopf M, Remaury A, Bockaert J, Manzoni OJ (2002) Endogenous cannabinoids mediate long-term synaptic depression in the nucleus accumbens. Proc Natl Acad Sci USA 99:8384–8388

Ronesi J, Gerdeman GL, Lovinger DM (2004) Disruption of endocannabinoid release and striatal long-term depression by postsynaptic blockade of endocannabinoid membrane transport. J Neurosci 24:1673–1679

Ryberg E, Larsson N, Sjogren S, Hjorth S, Hermansson N-O, Leonova J, Elebring T, Nilsson K, Drmota T, Greasley PJ (2007) The orphan receptor GPR55 is a novel cannabinoid receptor. Br J Pharmacol 152:1092–1101

Safo PK, Regehr WG (2005) Endocannabinoids control the induction of cerebellar LTD. Neuron 48:647–659

Sang N, Zhang J, Chen C (2006) PGE2 glycerol ester, a COX-2 oxidative metabolite of 2-arachidonoyl glycerol, modulates hippocampal inhibitory synaptic transmission in mouse hippocampal neurons. J Physiol (Lond) 572:735–745

Singla S, Kreitzer AC, Malenka RC (2007) Mechanisms for synapse specificity during striatal long-term depression. J Neurosci 27:5260–5264

Sjostrom PJ, Turrigiano GG, Nelson SB (2003) Neocortical LTD via coincident activation of presynaptic NMDA and cannabinoid receptors. Neuron 39:641–654

Stella N, Schweitzer P, Piomelli D (1997) A second endogenous cannabinoid that modulates long-term potentiation. Nature 388:773–778

Straiker A, Mackie K (2005) Depolarization-induced suppression of excitation in murine autaptic hippocampal neurones. J Physiol (Lond) 569:501–517

Sugiura T, Kondo S, Sukagawa A, Nakane S, Shinoda A, Itoh K, Yamashita A, Waku K (1995) 2-Arachidonoylglycerol: a possible endogenous cannabinoid receptor ligand in brain. Biochem Biophys Res Commun 215:89–97

Sullivan JM (2000) Cellular and molecular mechanisms underlying learning and memory impairments produced by cannabinoids. Learn Mem 7:132–139

Szabadits E, Cserep C, Ludanyi A, Katona I, Gracia-Llanes J, Freund TF, Nyiri G (2007) Hippocampal GABAergic synapses possess the molecular machinery for retrograde nitric oxide signaling. J Neurosci 27:8101–8111

Tzounopoulos T, Rubio ME, Keen JE, Trussell LO (2007) Coactivation of pre- and postsynaptic signaling mechanisms determines cell-specific spike-timing-dependent plasticity. Neuron 54: 291–301

Uchigashima M, Narushima M, Fukaya M, Katona I, Kano M, Watanabe M (2007) Subcellular arrangement of molecules for 2-arachidonoyl-glycerol-mediated retrograde signaling and its physiological contribution to synaptic modulation in the striatum. J Neurosci 27:3663–3676

Van Sickle MD, Duncan M, Kingsley PJ, Mouihate A, Urbani P, Mackie K, Stella N, Makriyannis A, Piomelli D, Davison JS, Marnett LJ, Di Marzo V, Pittman QJ, Patel KD, Sharkey KA (2005) Identification and functional characterization of brainstem cannabinoid CB2 receptors. Science 310:329–332

Varma N, Carlson GC, Ledent C, Alger BE (2001) Metabotropic glutamate receptors drive the endocannabinoid system in hippocampus. J Neurosci 21:RC188(1–5)

Varma N, Brager DH, Morishita W, Lenz RA, London B, Alger B (2002) Presynaptic factors in the regulation of DSI expression in hippocampus. Neuropharmacology 43:550–562

Wagner JJ, Alger BE (1996) Increased neuronal excitability during depolarization-induced suppression of inhibition in rat hippocampus. J Physiol (Lond) 495:107–112

Wang J, Zucker RS (2001) Photolysis-induced suppression of inhibition in rat hippocampal CA1 pyramidal neurons. J Physiol (Lond) 533:757–763

Wilson RI, Nicoll RA (2001) Endogenous cannabinoids mediate retrograde signalling at hippocampal synapses. Nature 410:588–592

Wilson RI, Kunos G, Nicoll RA (2001) Presynaptic specificity of endocannabinoid signaling in the hippocampus. Neuron 31:453–462

Yoshida T, Fukaya M, Uchigashima M, Miura E, Kamiya H, Kano M, Watanabe M (2006) Localization of diacylglycerol lipase-a around postsynaptic spine suggests close proximity between production site of an endocannabinoid, 2-arachidonoyl-glycerol, and presynaptic cannabinoid CB1 receptor. J Neurosci 26:4740–4751

Zhu PJ, Lovinger DM (2007) Persistent synaptic activity produces long-lasting enhancement of endocannabinoid modulation and alters long-term synaptic plasticity. J Neurophysiol 97:4386–4389

Zhuang S, Hampson RE, Deadwyler SA (2005) Behaviorally relevant endocannabinoid action in hippocampus: dependence on temporal summation of multiple inputs. Behav Pharmacol 16:463–471

Part IV
Future Directions

Chapter 9
Interneuron Pathophysiologies: Paths to Neurodevelopmental Disorders

Kathie L. Eagleson, Elizabeth A.D. Hammock, and Pat Levitt

9.1 Introduction

The mature mammalian cerebral cortex is characterized by its organization in to discrete areas that sub-serve higher sensory, motor and cognitive functions. Although pyramidal neurons provide the principal information outflow from a given cortical region, GABAergic interneurons have an essential role in providing inhibitory tone to local circuits, thereby regulating the firing rates and coordinating the final output of multiple pyramidal cells. More specifically, interneurons are able to modulate sensory gating and enhance discriminative information processing, for example, fine tuning sensory maps (Calford 2002; Hensch and Stryker 2004; Kaur et al. 2004) and optimizing executive functioning such as working memory (Rao et al. 2000; Constantinidis et al. 2002). As detailed in previous chapters, sub-classes of cortical interneurons have been identified based on electrophysiological, morphological and biochemical properties. While the relative ratio of excitatory to inhibitory neurons is constant across species and across most areas of the mature neocortex, the laminar distribution and relative number of different sub-populations of interneurons vary across discrete architectonic areas (Hendry et al. 1987; Hogan et al. 1992; Szabat et al. 1992; Alcántara and Ferrer 1994; DeFelipe et al. 1999; Gao et al. 1999, 2000; Hof et al. 1999; Cruikshank et al. 2001; Elston and Gonzalez-Albo 2003). This variation is thought to reflect regional differences in

---

K.L. Eagleson (✉)
Zilkha Neurogenetic Institute and the Department of Cell and Neurobiology, Keck School of Medicine at USC, 1501 San Pablo Street, Los Angeles, CA, 90089, USA
e-mail: keagleso@usc.edu

E.A.D. Hammock
Vanderbilt Kennedy Center for Research on Human Development and the Department of Pharmacology, Vanderbilt University Medical Center, 8114 Medical Research Building III, 465 21st Avenue South, Nashville, TN, 37203, USA

P. Levitt
Zilkha Neurogenetic Institute and the Department of Cell and Neurobiology, Keck School of Medicine at USC, 1501 San Pablo Street, Los Angeles, CA, 90089, USA

the organization and the function of local circuits. In addition, species differences in regional cytoarchitecture offer support for the specialization of function of local circuitry. Although developmental differences may exist between primate and non-primate species (Letinic et al. 2002), large numbers of interneurons relocate from their origin in the ganglionic eminence of the ventral telencephalon to their appropriate location in dorsal cerebral cortex (de Carlos et al. 1996; Anderson et al. 1997; Tamamaki et al. 1997; Lavdas et al. 1999; Letinic et al. 2002). Once in their final position in the cortex, interneurons undergo a prolonged maturation process that lasts long into the postnatal period and includes the neurochemical differentiation of these cells and the formation of GABAergic synapses (Blue and Parnavelas 1983; Miller 1986; Alcantara et al. 1993; Alcántara et al. 1993; Huang et al. 1999). This experience-dependent maturation process is critical for generating the appropriate GABAergic modulation of pyramidal cell function required for sensory processing, learning and memory, emotion regulation and cognitive function.

Given the importance of GABAergic circuitry for cortical function, one would predict that disruptions in the generation, migration and differentiation of cortical interneurons could have a profound effect on the level of excitability and the quality of information processing within the cortex, leading to significant functional deficits. Such deficits are the hallmark of many brain-based disorders. While some of these deficits are easy to discern, for example an increased incidence of spontaneous seizures potentially leading to epilepsy, other deficits associated with altered interneuron functioning may at first glance be less obvious. For example, there is evidence that abnormalities in cortical interneurons disrupt the organization of the minicolumn, the basic modular unit of physiological processing in the cortex (Mountcastle 1997), thus altering the quality of information integration that occurs. Reduced quality in turn affects multi-modal processing within the cortex that is critical for adaptive responses to unexpected stimuli and optimal functional performance, including motor output, attention, and emotional regulation. Although in this chapter we focus mainly on cortical interneurons, it should be noted that sub-cortical inhibitory circuits also are likely to be affected following alterations in interneuron development, given that GABAergic neurons generated in the ganglionic eminence populate structures throughout the forebrain. For example, sleep disturbances occur in many neurodevelopmental disorders (Malow 2004), and the circuits involved in sleep regulation may be disrupted leading to disturbances in the pattern of the sleep/wake cycle; pharmacological studies indicate that appropriate levels of inhibition are critical to the functioning of these circuits, as they can be modulated by $GABA_A$ receptor signaling [for review, see (Mohler 2007)].

Observations in mutant mice, which provide an opportunity to manipulate genetically specific histogenic events more readily than can be achieved through pharmacological treatments, have validated many of the predictions regarding interneuron pathophysiology and disrupted function. For example, mice lacking the *Dlx1* gene exhibit a reduction in the calretinin and somatostatin sub-populations of cortical and hippocampal interneurons by 1 month of age. This loss of interneurons is associated with reliable induction of seizures by mild stressors from the second post-natal month (Cobos et al. 2005). In *uPAR* null mice, a selective loss of parvalbumin-expressing interneurons from more anterior regions of the cortex and the somatostatin sub-population of interneurons from the hippocampus leads to increased sensitivity to convulsants, heightened anxiety, and a disruption of social interactions (Powell et al. 2003; Eagleson et al. 2005; Levitt 2005). In both examples, gene loss results in fewer cortical interneurons, which is then associated with heightened seizure susceptibility and alterations in emotional regulation.

9.2 Brain-Based Disorders

Based on the predictions outlined above, the clinical profiles of many brain-based developmental disorders, which often include deficits in information processing, increased incidence of spontaneous seizures and perturbations in the sleep–wake cycle, have led to the hypothesis that abnormal functioning of GABAergic interneurons is a key component in the underlying pathophysiology (Levitt et al. 2004; Lewis et al. 2005). However, direct evidence for interneuron dysfunction for most of these disorders is limited, and principally comes from postmortem studies demonstrating alterations in specific components of the GABAergic system or from the behavior-modulating effects of drugs that target the GABAergic system. It should be noted that the recent advances in imaging technology have begun to provide additional support for atypical interneuron functioning. In this section, we discuss the evidence that currently exists to support a role of interneuron pathology in three relatively common disorders, epilepsy, schizophrenia and autism spectrum disorder, as well as in some rarer syndromes and in generalized intellectual disability (Table 9.1).

9.2.1 Epilepsy

Epilepsy is one of the most common neurological disorders, affecting just under 1% of the population. Given that epilepsy reflects an upset in the balance between excitation and inhibition, it is not surprising that dysfunction of GABAergic interneurons has been strongly implicated in epilepsy. Many of the drugs currently used to treat epilepsy, including the benzodiazepines, target $GABA_A$ receptors, while antagonists of the $GABA_A$ receptor induce seizures. As noted above, mouse lines in which there is a disruption in interneuron development often display an increased susceptibility to seizures. Consistent with this, reductions in the number of interneurons and the expression of $GABA_A$ receptor sub-units are reliable findings in tissue resections from patients with epilepsy [for example, (de Lanerolle et al. 1989; DeFelipe 1999; Loup et al. 2006)]. As discussed in more detail in the

Table 9.1 Direct and indirect evidence for GABAergic interneuron dysfunction in select neurodevelopmental disorders

Disorder	Clinical profile (indirect)	Neuropathology (direct)	Pharmacology (direct)	Molecular (indirect)
Schizophrenia	Cognitive deficits including impaired working memory and executive function	Decreased GAD67 and parvalbumin mRNA expression; reduction in chandelier cartridges	Alteration in $GABA_A$ receptor sub-unit expression; increase in $GABA_A$ binding activity; reduced GABA transporter expression	NP
Autism Spectrum Disorder	Atypical sensory processing; increased incidence of seizures and sleep disorders; cognitive deficits involving specific modalities	Disruption in minicolumn organization; decreased expression of GAD65 and GAD67	Reduction in $GABA_A$ receptor binding; reduced expression of the β3 sub-unit of $GABA_A$ receptor	Genetic association with allelic variants of genes in the $GABA_A$ receptor cluster (α5, β3, γ3) and with *MET*
Epilepsy	Seizures	Decrease in interneurons	Decreased expression of $GABA_A$ receptor sub-units; $GABA_A$ receptor agonists relieve seizures	NP
Tuberous Sclerosis	>80% have epilepsy; deficits in executive function and attention; increased incidence of ASD and sleep disorders	Cortical tubers; alterations in interneuron markers	Alteration in $GABA_A$ receptor sub-unit expression within dysplastic cortex	NP
Fragile X	Hyper-responsiveness to sensory stimuli; increased incidence of seizures, ASD and sleep disorders	NP	NP	NP
Prader–Willi	Increased incidence of seizures and ASD; high co-morbidity with sleep disorders and compulsions	NP	Decreased $GABA_A$ receptor binding; topiramate modulates compulsive behaviors	Deletion or hypomorphic expression of $GABA_A$ receptor cluster (α5, β3, γ3)

Angleman	Abnormal EEG with >80% overt seizures; increased incidence of ASD and sleep disorders	NP	Deletion or hypomorphic expression of GABA$_A$ receptor cluster ($\alpha 5$, $\beta 3$, $\gamma 3$)
Rett	Gross cognitive impairment; increased incidence of seizures and ASD	NP	Misregulation of *BDNF* and *DLX5* genes
Intellectual disability – ARX	Increased incidence of seizures	Reduced expression of the $\beta 3$ sub-unit of GABA$_A$ receptor	Loss-of-function or hypomorphic ARX gene
Intellectual disability – SSADH	Increased incidence of seizures	Reduced expression of the $\beta 3$ sub-unit of GABA$_A$ receptor	NP
	Increase in GABA and GHB levels	NP	NP

NP not published

following sections there is an increased incidence of epilepsy in many brain-based disorders, supporting a role for interneuron dysfunction as a common mechanism across many disorders.

9.2.2 Schizophrenia

Schizophrenia affects approximately 1% of the population, with males and females affected equally. Among the clinical features of this disorder are positive symptoms (including hallucinations and delusions), negative symptoms (including flat affect and anhedonia), and cognitive deficits (for example, poor executive function, including impaired attention and working memory). Marked functional impairment resulting from these symptoms typically presents for the first time during the late adolescent/early adult period, which corresponds to the final stages of experience-dependent synapse pruning and a slowing of myelination of frontal and temporal fiber pathways.

A role for GABAergic neurons in the pathophysiology of schizophrenia was hypothesized over 30 years ago (Roberts 1972). Thus, multiple postmortem studies focused on this system, providing clear evidence for dysfunction within the GABAergic system in this disorder. In particular, the known deficits in working memory of patients with schizophrenia suggested that the prefrontal cortex would be an appropriate region on which to focus. As a consequence, a number of studies have now reported alterations in markers of GABAergic function in the prefrontal cortex of people with schizophrenia (Lewis and Hashimoto 2007), including decreases in GAD67 and parvalbumin mRNA expression (Akbarian et al. 1995; Volk et al. 2000; Reynolds and Beasley 2001; Hashimoto et al. 2003, 2008), increases and decreases in the expression of specific sub-units of the $GABA_A$ receptor (Huntsman et al. 1998; Ohnuma et al. 1999; Volk et al. 2002; Hashimoto et al. 2008), an increase in $GABA_A$ receptor binding activity (Benes et al. 1996b; Dean et al. 1999), a reduction in the expression of the GABA transporter (Ohnuma et al. 1999; Volk et al. 2001) and a reduction in the number of arrays of axon terminals of chandelier neurons (chandelier cartridges) (Woo et al. 1998; Pierri et al. 1999). In addition to alterations in several direct markers of GABA signaling, the levels of the TrkB receptor and its ligand, BDNF, which play a critical role in the differentiation of cortical interneurons (Marty et al. 1997; Rutherford et al. 1997), are reduced at both the protein and transcript level in the brains of patients with schizophrenia (Hashimoto et al. 2005). It should be noted that there is similar though less extensive evidence for GABAergic dysfunction in other brain regions in patients with this disorder, including the hippocampus (Benes et al. 1996a), temporal cortex (Reynolds et al. 2002; Deng and Huang 2006) and cerebellum (Fatemi et al. 2005). Taken together, these pathophysiological findings indicate that specific sub-populations of cortical interneurons are affected preferentially in schizophrenia, perhaps most interestingly, the chandelier cells that project to the axon initial segment of pyramidal cells. The chandelier interneurons are positioned to exert a powerful influence on the excitatory output of long projection neurons.

9.2.3 Autism Spectrum Disorder

Autism spectrum disorder (ASD) is characterized by deficits in social interactions, impairments in verbal and non-verbal communication, and restricted, repetitive, stereotyped patterns of behavior. Atypical sensory processing, including deficits in all five senses as well as vestibular and proprioceptive inputs, also has been reported in ASD, and some authors suggest that this should be considered a core feature of the disorder (Tecchio et al. 2003; Kern et al. 2006, 2007; Baker et al. 2008; Tomchek and Dunn 2007; Tommerdahl et al. 2007). In addition, there are several co-morbid conditions associated with sub-groups of individuals diagnosed with ASD, including an increased incidence of seizures (Tuchman and Rapin 2002; Canitano 2007), intellectual disability, and sleep disorders (Malow 2004). The most recent estimate of the prevalence of this disorder indicated that it could be as high as 1 in 155 children affected (CDC 2007), with the risk three to four times higher in males than in females.

Many aspects of the clinical profile of ASD indicate that atypical interneuron functioning may be involved (Hussman 2001; Dhossche et al. 2002; Rubenstein and Merzenich 2003; Levitt et al. 2004). In addition, Minshew and colleagues have proposed an alternative model based on the disordered complex information processing as a key component of ASD (Minshew et al. 2002; Williams et al. 2006). Both the interneuron hypothesis and the Minshew model are consistent with the reports of disruption in minicolumn organization in the cortex of patients with ASD, particularly as this anatomical feature is thought to be modulated by interneuron function (Casanova et al. 2002a, b; Buxhoeveden et al. 2006). Human genetic and postmortem findings show an association of ASD with the gene encoding the MET tyrosine kinase receptor (Campbell et al. 2006, 2007), which is implicated in interneuron development by studies in the mouse (Powell et al. 2001, 2003; Eagleson et al. 2005). Moreover, allelic variants in the gene encoding the β3 sub-unit of the $GABA_A$ receptor have been associated with ASD (McCauley et al. 2004). Additional studies implicate genetic variants and chromosomal inversions in genes encoding other $GABA_A$ receptor sub-units (Ma et al. 2005; Ashley-Koch et al. 2006; Vincent et al. 2006). Unlike schizophrenia and epilepsy, however, there is little direct neuroanatomical evidence for an involvement of cortical interneurons in the pathology of ASD and, thus far, the sample sizes in many postmortem studies are small and the studies generally have not been replicated. Nonetheless, some intriguing observations have been reported. Within the hippocampus, there is a specific reduction in $GABA_A$ receptors in ASD, with serotonergic, cholinergic and glutamatergic receptors largely intact (Blatt et al. 2001), although it should be noted that a more recent study demonstrated alterations in discrete sub-units of the cholinergic receptor in the cerebral cortex and cerebellum (Martin-Ruiz et al. 2004). A reduction in the levels of both GAD65 and GAD67 has been reported in the parietal cortex and cerebellum in ASD (Fatemi et al. 2002), as well as reduced frontal cortex expression of the β3 sub-unit of the $GABA_A$ receptor (Samaco et al. 2005). Finally, reduced levels of GABA in peripheral platelets (Rolf et al. 1993) as well as elevated GABA levels in plasma (Dhossche et al. 2002), have been reported; however, the functional significance of these findings with respect to the brain is unclear. Interestingly, several neurodevelopmental disorders are associated with an increased incidence of ASD [for review, (Zafeiriou et al. 2007)], including tuberous sclerosis, Fragile X syndrome, Rett syndrome, and both Angelman and Prader–Willi syndromes. Disruptions in GABAergic interneurons have also been implicated in these syndromes (see below for details), suggesting potential overlapping pathophysiologies across multiple brain-based disorders of developmental etiology.

9.2.4 Other Developmental Disorders

9.2.4.1 Tuberous Sclerosis

Tuberous sclerosis (TSC) is a rare genetic disorder, occurring in approximately 1:10,000 people with males and females affected equally, that causes benign tumors in the brain, as well as in other organs. This disorder results from mutations in one of two tumor suppressor genes, *TSC1* or *TSC2* (ECTS 1993; van Slegtenhorst et al. 1997). Epilepsy occurs in approximately 80% of patients with TSC (Joinson et al. 2003) and often presents within the first year of life, with other behavioral and cognitive deficits becoming apparent with development. These include attentional and executive memory deficits, as well as depression, anxiety, aggression, and disturbances in sleep patterns (de Vries et al. 2005). Approximately 50% of individuals with TSC have an intellectual disability (IQ < 70) (Joinson et al. 2003) and the prevalence of ASD is higher than in the typical population, although estimates vary widely from around 16% to over 65%, depending on the study (Smalley 1998; Wong 2006).

Almost all aspects of the clinical profile of TSC are indicative of a disturbance in GABAergic functioning. One of the characteristics of TSC pathology is the presence of tubers within the cerebral cortex [for review (Curatolo et al. 2002)]. These structures, in which lamination is disorganized and many cells have abnormal morphologies, are the sites of seizure initiation. As many patients with TSC are refractory to pharmacological therapy, surgical resection of cortical tubers may be required to enable seizure control (Shields 2004), providing a source of tissue for analysis. Thus far, however, only two published studies, both involving small sample sizes, have focused directly on the GABAergic system in TSC, and alterations in the expression of interneuron markers and $GABA_A$ receptor sub-units were noted within the dysplastic cortex (White et al. 2001; Valencia et al. 2006).

9.2.4.2 Fragile X

Fragile X is the most common single-gene inherited form of intellectual disability and is estimated to occur in approximately 1:4,000 births, with about 1.5 times as many males affected. The syndrome arises due to disruption of the Fragile X mental retardation (FMR1) gene, located on the X chromosome, by a trinucleotide repeat expansion (Verkerk et al. 1991), which results in the absence of the FMR protein. Individuals with this syndrome display mild to severe cognitive impairment and typically demonstrate a neurobehavioral profile that includes hyper-responsiveness to sensory stimuli, hyperactivity and impulsivity. In addition, over 20% of people with Fragile X have overt seizures (Musumeci et al. 1999) and, as with many neurodevelopmental disorders, disruptions in sleep patterns are common. Finally, about 25% of those with Fragile X display traits that are also used to diagnose ASD (Bailey et al. 1998).

Among the clinical features associated with this disorder, hyper-responsiveness to sensory stimuli together with the increased incidence of seizures and ASD are particularly suggestive of an interneuron dysfunction. Thus far, however, components of the GABAergic system have not been examined in human postmortem material. It is interesting to note, however, that in species as diverse as mouse and fly, mutations in *FMR1* lead to a reduction in the expression of various $GABA_A$ receptor sub-units (D'Hulst et al. 2006; Gantois et al. 2006). Given the highly conserved nature of FMR1 protein structure and function, these data suggest a potential evolutionarily-conserved relationship between a functional FMR1 protein and GABAergic function. Similarly, electrophysiological studies in the subiculum of *FMR1* knockout mice suggest that there is a disruption of $GABA_A$ receptor-mediated function in this structure (D'Antuono et al. 2003).

9.2.4.3 15q11–q13 and Gene Regulatory Disorders: Prader–Willi, Angelman, and Rett Syndromes

Prader–Willi syndrome (PWS) and Angelman syndrome (AS) result from abnormalities, including deletions, associated with the chromosomal region 15q11–q13 that lead to disruption in the expression of genes from this region. The overall prevalence of each syndrome is approximately 1:12,000–1:15,000, with males and females affected equally. The genetic distinction between the two disorders is that PWS involves gene abnormalities on the paternally donated chromosome whereas gene abnormalities on the maternally inherited chromosome gives rise to AS. There is now evidence that deficiency of the maternally inherited E6-AP ubitquitin protein ligase (*UBE3A*) gene is both necessary and sufficient to cause AS (Kishino et al. 1997; Matsuura et al. 1997). Also of note with respect to interneuron dysfunction, there is a cluster of three $GABA_A$ receptor sub-units ($\alpha 5$, $\beta 3$ and $\gamma 3$) in this chromosomal region that are deleted or hypomorphic in most people with PWS or AS (Saitoh et al. 1994). Although the same chromosomal region is affected, PWS and AS display distinct, yet partially overlapping, clinical profiles, highlighting the role of maternal imprinting on phenotypic outcome. PWS is characterized by intellectual disability, infantile hypotonia and poor suck reflex, and delayed sexual development, with high co-morbidity for depression, obsessions and compulsions, self-injurious behavior (usually in the form of self-inflicted skin picking) and sleep disorders (Holm et al. 1993). Perhaps the most striking feature of PWS is an intense preoccupation with food, following an early failure-to-thrive period, manifesting as incessant food-seeking and a lack of satiation that often results in obesity. In contrast, AS is characterized by hyperactivity, stereotypies and sleep disorders, as well as severe intellectual disability and the absence of speech, although receptive language and non-verbal communication can be relatively preserved. Distinctive EEG abnormalities are seen in over 90% of patients with AS, with more than 80% exhibiting overt seizures (Valente et al. 2006; Pelc et al. 2008).

Rett Syndrome (RTT) is a rare (1:15,000) X-linked dominant disorder expressed almost exclusively in females. In most individuals, RTT results from mutations in the gene encoding methyl-CpG-binding protein 2 (MeCP2), which serves as a transcriptional repressor of imprinted regions of Chromosome 15 and select target genes throughout the genome (Amir et al. 1999; Caballero and Hendrich 2005; Horike et al. 2005; Samaco et al. 2005). Among others, MeCP2 regulates expression of *UBE3A*, which is the gene disrupted in AS, *DLX5*, which has been implicated in GABAergic differentiation, and *bdnf*, whose gene product regulates interneuron and synaptic development. Children with RTT regress after achieving typical motor and speech milestones between 6 and 18 months. By 4–7 years of age, gross motor and cognitive impairments, loss of speech, and reduced growth trajectories of body and brain are evident. Like AS and PWS, a large proportion of children with RTT exhibits seizures and co-occurring ASD. The overlap in clinical symptoms between the three disorders, in particular between AS and RTT, suggests that there may be some shared pathophysiology.

In particular, the extremely high co-morbidity of seizures with RTT and AS is indicative of altered interneuron function. Indeed, specific hypotheses have been proposed regarding GABAergic dysfunction, including altered cortical inhibitory circuits, in the underlying pathophysiology of AS (Dan and Boyd 2003). Consistent with this, there is reduced expression of the β3 sub-unit of the $GABA_A$ receptor in frontal cortex of subjects with AS and RTT (Samaco et al. 2005). The GABAergic system also has been implicated in PWS, although direct evidence is limited and includes studies with a small sample size. Moreover, although many of the clinical features of PWS suggest a primary hypothalamic dysfunction, regions of the frontal and temporal cortices are also involved in the emotional responses to satiety and hunger (Tataranni et al. 1999). A recent study using positron emission tomography (PET) demonstrated a reduction in the binding of [^{11}C]flumazenil in frontal and temporal cortical regions in PWS patients, reflecting an alteration in the sub-unit composition or number of $GABA_A$ receptors (Lucignani et al. 2004). It has been reported that topiramate, a drug that influences GABAergic signaling, is able to modulate the stereotypic and compulsive behaviors observed in PWS (Shapira et al. 2002; Smathers et al. 2003). Finally, there is an increase in the levels of GABA found in the plasma of patients with both PWS and AS (Ebert et al. 1997), although, as for ASD, the functional relevance to the brain remains unclear.

9.2.4.4 Intellectual Disability

Intellectual disability (ID), currently defined as an IQ below 70, impairment in adaptive functioning, and an age of onset prior to 18 years, can result from a variety of genetic and environmental insults, although the cause is never identified in approximately half of this population. It is estimated that between 1 and 3% of the population has an ID, including those that are co-morbid for the disorders outlined earlier. The prevalence of epilepsy in the population of people with ID is significantly increased above the general population, although estimates of overall prevalence vary depending on the study (Goulden et al. 1991; Bowley and Kerr 2000; Morgan et al. 2003). The specific form of epilepsy, as well as the frequency of seizures, varies across the population and likely reflects the underlying etiology of the intellectual impairment (Beavis et al. 2007). The increased incidence of epilepsy in ID indicates that interneuron dysfunction may be a common pathophysiology contributing to ID, regardless of the specific cause. Here, we highlight two known etiologies of ID that are related directly to interneuron function. The first involves the Aristaless-related homeobox gene (*Arx*), which is involved in GABAergic neuron development in species as diverse as worm and mouse (Kitamura et al. 2002; Melkman and Sengupta 2005). For example, in the mouse, this gene is important in the migration of cortical interneurons to the cerebral cortex (Kitamura et al. 2002). In humans, mutations in *ARX*, which can lead to a loss-of-function or a hypomorphic state of expression, have a variety of clinical manifestations, including ID and epilepsy [reviewed in (Sherr 2003)]. The second involves disorders of GABA metabolism, the most common of which is a succinic semialdehyde dehydrogenase (SSADH) deficiency that affects GABA degradation in the central nervous system. This is a very rare disorder that results in an increase in GABA and gamma-hydroxybutyric acid (GHB) levels in the brain. About half the people with this disorder display mild to moderate intellectual disability, particularly involving language deficits, in addition to seizures, motor delay and hallucinations (Pearl et al. 2003). The increase in brain GABA content in this rare disorder is in contrast with the other disorders outlined earlier that show decreases in GABA function and signaling. While the mechanism is not specifically known, this example serves to underscore the importance of homeostatic balance in maintaining functional integrity of the two major neurotransmitters in the brain, namely GABA and glutamate.

9.3 Conclusions

Disorders of neurodevelopmental etiology are diverse in their onset and manifestation, yet they appear to have in common fundamental disturbances in cortical GABAergic function. GABA neurotransmission is essential for the experience-dependent maturation of sensory representations in the brain, and in the processing of complex information through the role of interneurons as coincidence detectors and regulators of output synchrony. Genetic disruptions of fundamental developmental events that regulate interneuron development are, therefore, likely to establish vulnerabilities that are further exacerbated by atypical experience-dependent maturation. It should be emphasized that although the onset of many of these neurodevelopmental disorders is defined clinically as the first manifestation of a disruption in discrete behavioral and/or cognitive abilities, with many disorders being diagnosed within the first 2–3 years of life, suggestions of atypical development may be observed earlier. Thus, the age of onset often reflects the developmental emergence of the specific behaviors/abilities, as well as the maturational state of the underlying circuitry. This occurs even for disorders with a later onset, such as schizophrenia, where the first overt signs of the disorder correlate with the final stages of maturation of prefrontal cortical circuitry. Future emphasis on understanding gene-environment relationships in mediating interneuron development will better inform disease etiologies.

Acknowledgments This work is supported by NICHD P30 grant HD15052, the Marino Autism Research Institute and NIMH grant MH067842.

References

Akbarian S, Huntsman MM, Kim JJ, Tafazzoli A, Potkin SG, Bunney WE Jr, Jones EG (1995) GABAA receptor subunit gene expression in human prefrontal cortex: comparison of schizophrenics and controls. Cereb Cortex 5:550–560

Alcántara S, Ferrer I (1994) Postnatal development of parvalbumin immunoreactivity in the cerebral cortex of the cat. J Comp Neurol 348:133–149

Alcantara S, Ferrer I, Soriano E (1993) Postnatal development of parvalbumin and calbindin D28K immunoreactivities in the cerebral cortex of the rat. Anat Embryol (Berl) 188:63–73

Alcántara S, Ferrer I, Soriano E (1993) Postnatal development of parvalbumin and calbindin D28K immunoreactivities in the cerebral cortex of the rat. Anat Embryol (Berl) 188:63–73

Amir RE, Van den Veyver IB, Wan M, Tran CQ, Francke U, Zoghbi HY (1999) Rett syndrome is caused by mutations in X-linked MECP2, encoding methyl-CpG-binding protein 2. Nat Genet 23:185–188

Anderson SA, Eisenstat DD, Shi L, Rubenstein JL (1997) Interneuron migration from basal forebrain to neocortex: dependence on Dlx genes. Science 278:474–476

Ashley-Koch AE, Mei H, Jaworski J, Ma DQ, Ritchie MD, Menold MM, Delong GR, Abramson RK, Wright HH, Hussman JP, Cuccaro ML, Gilbert JR, Martin ER, Pericak-Vance MA (2006) An analysis paradigm for investigating multi-locus effects in complex disease: examination of three GABA receptor subunit genes on 15q11–q13 as risk factors for autistic disorder. Ann Hum Genet 70:281–292

Bailey DB Jr, Mesibov GB, Hatton DD, Clark RD, Roberts JE, Mayhew L (1998) Autistic behavior in young boys with fragile X syndrome. J Autism Dev Disord 28:499–508

Baker AE, Lane A, Angley MT, Young RL (2008) The relationship between sensory processing patterns and behavioural responsiveness in autistic disorder: a pilot study. J Autism Dev Disord 38(5):867–875

Beavis J, Kerr M, Marson AG (2007) Pharmacological interventions for epilepsy in people with intellectual disabilities. Cochrane Database Syst Rev CD005399

Benes FM, Khan Y, Vincent SL, Wickramasinghe R (1996a) Differences in the subregional and cellular distribution of GABAA receptor binding in the hippocampal formation of schizophrenic brain. Synapse 22:338–349

Benes FM, Vincent SL, Marie A, Khan Y (1996b) Up-regulation of GABAA receptor binding on neurons of the prefrontal cortex in schizophrenic subjects. Neuroscience 75:1021–1031

Blatt GJ, Fitzgerald CM, Guptill JT, Booker AB, Kemper TL, Bauman ML (2001) Density and distribution of hippocampal neurotransmitter receptors in autism: an autoradiographic study. J Autism Dev Disord 31:537–543

Blue ME, Parnavelas JG (1983) The formation and maturation of synapses in the visual cortex of the rat. II. Quantitative analysis. J Neurocytol 12:697–712

Bowley C, Kerr M (2000) Epilepsy and intellectual disability. J Intellect Disabil Res 44(Pt 5):529–543

Buxhoeveden DP, Semendeferi K, Buckwalter J, Schenker N, Switzer R, Courchesne E (2006) Reduced minicolumns in the frontal cortex of patients with autism. Neuropathol Appl Neurobiol 32:483–491

Caballero IM, Hendrich B (2005) MeCP2 in neurons: closing in on the causes of Rett syndrome. Hum Mol Genet 14 Spec No 1:R19–R26

Calford MB (2002) Dynamic representational plasticity in sensory cortex. Neuroscience 111:709–738

Campbell DB, Sutcliffe JS, Ebert PJ, Militerni R, Bravaccio C, Trillo S, Elia M, Schneider C, Melmed R, Sacco R, Persico AM, Levitt P (2006) A genetic variant that disrupts MET transcription is associated with autism. Proc Natl Acad Sci USA 103:16834–16839

Campbell DB, D'Oronzio R, Garbett K, Ebert PJ, Mirnics K, Levitt P, Persico AM (2007) Disruption of cerebral cortex MET signaling in autism spectrum disorder. Ann Neurol 62:243–250

Canitano R (2007) Epilepsy in autism spectrum disorders. Eur Child Adolesc Psychiatry 16:61–66

Casanova MF, Buxhoeveden DP, Brown C (2002a) Clinical and macroscopic correlates of minicolumnar pathology in autism. J Child Neurol 17:692–695

Casanova MF, Buxhoeveden DP, Switala AE, Roy E (2002b) Minicolumnar pathology in autism. Neurology 58:428–432

CDC (2007) Prevalence of autism spectrum disorders – Autism and Developmental Disabilities Monitoring Network. MMWR 56 (SS-1)

Cobos I, Calcagnotto ME, Vilaythong AJ, Thwin MT, Noebels JL, Baraban SC, Rubenstein JL (2005) Mice lacking Dlx1 show subtype-specific loss of interneurons, reduced inhibition and epilespy. Nat Neurosci 8:1059–1068

Constantinidis C, Williams GV, Goldman-Rakic PS (2002) A role for inhibition in shaping the temporal flow of information in prefrontal cortex. Nat Neurosci 5:175–180

Cruikshank SJ, Killackey HP, Metherate R (2001) Parvalbumin and calbindin are differentially distributed within primary and secondary subregions of the mouse auditory forebrain. Neuroscience 105:553–569

Curatolo P, Verdecchia M, Bombardieri R (2002) Tuberous sclerosis complex: a review of neurological aspects. Eur J Paediatr Neurol 6:15–23

Dan B, Boyd SG (2003) Angelman syndrome reviewed from a neurophysiological perspective. The UBE3A-GABRB3 hypothesis. Neuropediatrics 34:169–176

D'Antuono M, Merlo D, Avoli M (2003) Involvement of cholinergic and gabaergic systems in the fragile X knockout mice. Neuroscience 119:9–13

de Carlos JA, Lopez-Mascaraque L, Valverde F (1996) Dynamics of cell migration from the lateral ganglionic eminence in the rat. J Neurosci 16:6146–6156

de Lanerolle NC, Kim JH, Robbins RJ, Spencer DD (1989) Hippocampal interneuron loss and plasticity in human temporal lobe epilepsy. Brain Res 495:387–395

de Vries P, Humphrey A, McCartney D, Prather P, Bolton P, Hunt A (2005) Consensus clinical guidelines for the assessment of cognitive and behavioural problems in Tuberous Sclerosis. Eur Child Adolesc Psychiatry 14:183–190

Dean B, Hussain T, Hayes W, Scarr E, Kitsoulis S, Hill C, Opeskin K, Copolov DL (1999) Changes in serotonin2A and GABA(A) receptors in schizophrenia: studies on the human dorsolateral prefrontal cortex. J Neurochem 72:1593–1599

DeFelipe J (1999) Chandelier cells and epilepsy. Brain 122(Pt 10):1807–1822

DeFelipe J, Gonzalez-Albo MC, Del Rio MR, Elston GN (1999) Distribution and patterns of connectivity of interneurons containing calbindin, calretinin, and parvalbumin in visual areas of the occipital and temporal lobes of the macaque monkey. J Comp Neurol 412:515–526

Deng C, Huang XF (2006) Increased density of GABAA receptors in the superior temporal gyrus in schizophrenia. Exp Brain Res 168:587–590

Dhossche D, Applegate H, Abraham A, Maertens P, Bland L, Bencsath A, Martinez J (2002) Elevated plasma gamma-aminobutyric acid (GABA) levels in autistic youngsters: stimulus for a GABA hypothesis of autism. Med Sci Monit 8:PR1–PR6

D'Hulst C, De Geest N, Reeve SP, Van Dam D, De Deyn PP, Hassan BA, Kooy RF (2006) Decreased expression of the GABAA receptor in fragile X syndrome. Brain Res 1121:238–245

Eagleson KL, Bonnin A, Levitt P (2005) Region- and age-specific deficits in gamma-aminobutyric acidergic neuron development in the telencephalon of the uPAR(-/-) mouse. J Comp Neurol 489:449–466

Ebert MH, Schmidt DE, Thompson T, Butler MG (1997) Elevated plasma gamma-aminobutyric acid (GABA) levels in individuals with either Prader-Willi syndrome or Angelman syndrome. J Neuropsychiatry Clin Neurosci 9:75–80

ECTS C (1993) Identification and characterization of the tuberous sclerosis gene on chromosome 16. Cell 75:1305–1315

Elston GN, Gonzalez-Albo MC (2003) Parvalbumin-, calbindin-, and calretinin-immunoreactive neurons in the prefrontal cortex of the owl monkey (Aotus trivirgatus): a standardized quantitative comparison with sensory and motor areas. Brain Behav Evol 62:19–30

Fatemi SH, Halt AR, Stary JM, Kanodia R, Schulz SC, Realmuto GR (2002) Glutamic acid decarboxylase 65 and 67 kDa proteins are reduced in autistic parietal and cerebellar cortices. Biol Psychiatry 52:805–810

Fatemi SH, Stary JM, Earle JA, Araghi-Niknam M, Eagan E (2005) GABAergic dysfunction in schizophrenia and mood disorders as reflected by decreased levels of glutamic acid decarboxylase 65 and 67 kDa and Reelin proteins in cerebellum. Schizophr Res 72:109–122

Gantois I, Vandesompele J, Speleman F, Reyniers E, D'Hooge R, Severijnen LA, Willemsen R, Tassone F, Kooy RF (2006) Expression profiling suggests underexpression of the GABA(A) receptor subunit delta in the fragile X knockout mouse model. Neurobiol Dis 21:346–357

Gao WJ, Newman DE, Wormington AB, Pallas SL (1999) Development of inhibitory circuitry in visual and auditory cortex of postnatal ferrets: immunocytochemical localization of GABAergic neurons. J Comp Neurol 409:261–273

Gao WJ, Wormington AB, Newman DE, Pallas SL (2000) Development of inhibitory circuitry in visual and auditory cortex of postnatal ferrets: immunocytochemical localization of calbindin- and parvalbumin-containing neurons. J Comp Neurol 422:140–157

Goulden KJ, Shinnar S, Koller H, Katz M, Richardson SA (1991) Epilepsy in children with mental retardation: a cohort study. Epilepsia 32:690–697

Hashimoto T, Volk DW, Eggan SM, Mirnics K, Pierri JN, Sun Z, Sampson AR, Lewis DA (2003) Gene expression deficits in a subclass of GABA neurons in the prefrontal cortex of subjects with schizophrenia. J Neurosci 23:6315–6326

Hashimoto T, Bergen SE, Nguyen QL, Xu B, Monteggia LM, Pierri JN, Sun Z, Sampson AR, Lewis DA (2005) Relationship of brain-derived neurotrophic factor and its receptor TrkB to altered inhibitory prefrontal circuitry in schizophrenia. J Neurosci 25:372–383

Hashimoto T, Arion D, Unger T, Maldonado-Aviles JG, Morris HM, Volk DW, Mirnics K, Lewis DA (2008) Alterations in GABA-related transcriptome in the dorsolateral prefrontal cortex of subjects with schizophrenia. Mol Psychiatry 13(2):147–161

Hendry SH, Schwark HD, Jones EG, Yan J (1987) Numbers and proportions of GABA-immunoreactive neurons in different areas of monkey cerebral cortex. J Neurosci 7:1503–1519

Hensch TK, Stryker MP (2004) Columnar architecture sculpted by GABA circuits in developing cat visual cortex. Science 303:1678–1681

Hof PR, Glezer II, Conde F, Flagg RA, Rubin MB, Nimchinsky EA, Vogt Weisenhorn DM (1999) Cellular distribution of the calcium-binding proteins parvalbumin, calbindin, and calretinin in the neocortex of mammals: phylogenetic and developmental patterns. J Chem Neuroanat 16:77–116

Hogan D, Terwilleger ER, Berman NE (1992) Development of subpopulations of GABAergic neurons in cat visual cortical areas. Neuroreport 3:1069–1072

Holm VA, Cassidy SB, Butler MG, Hanchett JM, Greenswag LR, Whitman BY, Greenberg F (1993) Prader-Willi syndrome: consensus diagnostic criteria. Pediatrics 91:398–402

Horike S, Cai S, Miyano M, Cheng JF, Kohwi-Shigematsu T (2005) Loss of silent-chromatin looping and impaired imprinting of DLX5 in Rett syndrome. Nat Genet 37:31–40

Huang ZJ, Kirkwood A, Pizzorusso T, Porciatti V, Morales B, Bear MF, Maffei L, Tonegawa S (1999) BDNF regulates the maturation of inhibition and the critical period of plasticity in mouse visual cortex. Cell 98:739–755

Huntsman MM, Tran BV, Potkin SG, Bunney WE Jr, Jones EG (1998) Altered ratios of alternatively spliced long and short gamma2 subunit mRNAs of the gamma-amino butyrate type A receptor in prefrontal cortex of schizophrenics. Proc Natl Acad Sci USA 95:15066–15071

Hussman JP (2001) Suppressed GABAergic inhibition as a common factor in suspected etiologies of autism. J Autism Dev Disord 31:247–248

Joinson C, O'Callaghan FJ, Osborne JP, Martyn C, Harris T, Bolton PF (2003) Learning disability and epilepsy in an epidemiological sample of individuals with tuberous sclerosis complex. Psychol Med 33:335–344

Kaur S, Lazar R, Metherate R (2004) Intracortical pathways determine breadth of subthreshold frequency receptive fields in primary auditory cortex. J Neurophysiol 91:2551–2567

Kern JK, Trivedi MH, Garver CR, Grannemann BD, Andrews AA, Savla JS, Johnson DG, Mehta JA, Schroeder JL (2006) The pattern of sensory processing abnormalities in autism. Autism 10:480–494

Kern JK, Trivedi MH, Grannemann BD, Garver CR, Johnson DG, Andrews AA, Savla JS, Mehta JA, Schroeder JL (2007) Sensory correlations in autism. Autism 11:123–134

Kishino T, Lalande M, Wagstaff J (1997) UBE3A/E6-AP mutations cause Angelman syndrome. Nat Genet 15:70–73

Kitamura K, Yanazawa M, Sugiyama N, Miura H, Iizuka-Kogo A, Kusaka M, Omichi K, Suzuki R, Kato-Fukui Y, Kamiirisa K, Matsuo M, Kamijo S, Kasahara M, Yoshioka H, Ogata T, Fukuda T, Kondo I, Kato M, Dobyns WB, Yokoyama M, Morohashi K (2002) Mutation of ARX causes abnormal development of forebrain and testes in mice and X-linked lissencephaly with abnormal genitalia in humans. Nat Genet 32:359–369

Lavdas AA, Grigoriou M, Pachnis V, Parnavelas JG (1999) The medial ganglionic eminence gives rise to a population of early neurons in the developing cerebral cortex. J Neurosci 19:7881–7888

Letinic K, Zoncu R, Rakic P (2002) Origin of GABAergic neurons in the human neocortex. Nature 417:645–649

Levitt P (2005) Disruption of interneuron development. Epilepsia 46(Suppl 7):22–28

Levitt P, Eagleson KL, Powell EM (2004) Regulation of neocortical interneuron development and the implications for neurodevelopmental disorders. Trends Neurosci 27:400–406

Lewis DA, Hashimoto T (2007) Deciphering the disease process of schizophrenia: the contribution of cortical gaba neurons. Int Rev Neurobiol 78:109–131

Lewis DA, Hashimoto T, Volk DW (2005) Cortical inhibitory neurons and schizophrenia. Nat Rev Neurosci 6:312–324

Loup F, Picard F, Andre VM, Kehrli P, Yonekawa Y, Wieser HG, Fritschy JM (2006) Altered expression of alpha3-containing GABAA receptors in the neocortex of patients with focal epilepsy. Brain 129:3277–3289

Lucignani G, Panzacchi A, Bosio L, Moresco RM, Ravasi L, Coppa I, Chiumello G, Frey K, Koeppe R, Fazio F (2004) GABA A receptor abnormalities in Prader-Willi syndrome assessed with positron emission tomography and [11C]flumazenil. Neuroimage 22:22–28

Ma DQ, Whitehead PL, Menold MM, Martin ER, Ashley-Koch AE, Mei H, Ritchie MD, Delong GR, Abramson RK, Wright HH, Cuccaro ML, Hussman JP, Gilbert JR, Pericak-Vance MA (2005) Identification of significant association and gene-gene interaction of GABA receptor subunit genes in autism. Am J Hum Genet 77:377–388

Malow BA (2004) Sleep disorders, epilepsy, and autism. Ment Retard Dev Disabil Res Rev 10:122–125

Martin-Ruiz CM, Lee M, Perry RH, Baumann M, Court JA, Perry EK (2004) Molecular analysis of nicotinic receptor expression in autism. Brain Res Mol Brain Res 123:81–90

Marty S, Berzaghi Mda P, Berninger B (1997) Neurotrophins and activity-dependent plasticity of cortical interneurons. Trends Neurosci 20:198–202

Matsuura T, Sutcliffe JS, Fang P, Galjaard RJ, Jiang YH, Benton CS, Rommens JM, Beaudet AL (1997) De novo truncating mutations in E6-AP ubiquitin-protein ligase gene (UBE3A) in Angelman syndrome. Nat Genet 15:74–77

McCauley JL, Olson LM, Delahanty R, Amin T, Nurmi EL, Organ EL, Jacobs MM, Folstein SE, Haines JL, Sutcliffe JS (2004) A linkage disequilibrium map of the 1-Mb 15q12 GABA(A) receptor subunit cluster and association to autism. Am J Med Genet B Neuropsychiatr Genet 131:51–59

Melkman T, Sengupta P (2005) Regulation of chemosensory and GABAergic motor neuron development by the C. elegans Aristaless/Arx homolog alr-1. Development 132:1935–1949

Miller MW (1986) Maturation of rat visual cortex. III. Postnatal morphogenesis and synaptogenesis of local circuit neurons. Brain Res 390:271–285

Minshew NJ, Sweeney J, Luna B (2002) Autism as a selective disorder of complex information processing and underdevelopment of neocortical systems. Mol Psychiatry 7(Suppl 2):S14–S15

Mohler H (2007) Molecular regulation of cognitive functions and developmental plasticity: impact of GABAA receptors. J Neurochem 102:1–12

Morgan CL, Baxter H, Kerr MP (2003) Prevalence of epilepsy and associated health service utilization and mortality among patients with intellectual disability. Am J Ment Retard 108:293–300

Mountcastle VB (1997) The columnar organization of the neocortex. Brain 120(Pt 4):701–722

Musumeci SA, Hagerman RJ, Ferri R, Bosco P, Dalla Bernardina B, Tassinari CA, De Sarro GB, Elia M (1999) Epilepsy and EEG findings in males with fragile X syndrome. Epilepsia 40:1092–1099

Ohnuma T, Augood SJ, Arai H, McKenna PJ, Emson PC (1999) Measurement of GABAergic parameters in the prefrontal cortex in schizophrenia: focus on GABA content, GABA(A) receptor alpha-1 subunit messenger RNA and human GABA transporter-1 (HGAT-1) messenger RNA expression. Neuroscience 93:441–448

Pearl PL, Novotny EJ, Acosta MT, Jakobs C, Gibson KM (2003) Succinic semialdehyde dehydrogenase deficiency in children and adults. Ann Neurol 54(Suppl 6):S73–S80

Pelc K, Boyd SG, Cheron G, Dan B (2008) Epilepsy in Angelman syndrome. Seizure 17(3):211–217

Pierri JN, Chaudry AS, Woo TU, Lewis DA (1999) Alterations in chandelier neuron axon terminals in the prefrontal cortex of schizophrenic subjects. Am J Psychiatry 156:1709–1719

Powell EM, Mars WM, Levitt P (2001) Hepatocyte growth factor/scatter factor is a motogen for interneurons migrating from the ventral to dorsal telencephalon. Neuron 30:79–89

Powell EM, Campbell DB, Stanwood GD, Davis C, Noebels JL, Levitt P (2003) Genetic disruption of cortical interneuron development causes region- and GABA cell type-specific deficits, epilepsy, and behavioral dysfunction. J Neurosci 23:622–631

Rao SG, Williams GV, Goldman-Rakic PS (2000) Destruction and creation of spatial tuning by disinhibition: GABA(A) blockade of prefrontal cortical neurons engaged by working memory. J Neurosci 20:485–494

Reynolds GP, Beasley CL (2001) GABAergic neuronal subtypes in the human frontal cortex – development and deficits in schizophrenia. J Chem Neuroanat 22:95–100

Reynolds GP, Beasley CL, Zhang ZJ (2002) Understanding the neurotransmitter pathology of schizophrenia: selective deficits of subtypes of cortical GABAergic neurons. J Neural Transm 109:881–889

Roberts E (1972) Prospects for research on schizophrenia. An hypotheses suggesting that there is a defect in the GABA system in schizophrenia. Neurosci Res Program Bull 10:468–482

Rolf LH, Haarmann FY, Grotemeyer KH, Kehrer H (1993) Serotonin and amino acid content in platelets of autistic children. Acta Psychiatr Scand 87:312–316

Rubenstein JL, Merzenich MM (2003) Model of autism: increased ratio of excitation/inhibition in key neural systems. Genes Brain Behav 2:255–267

Rutherford LC, DeWan A, Lauer HM, Turrigiano GG (1997) Brain-derived neurotrophic factor mediates the activity-dependent regulation of inhibition in neocortical cultures. J Neurosci 17:4527–4535

Saitoh S, Harada N, Jinno Y, Hashimoto K, Imaizumi K, Kuroki Y, Fukushima Y, Sugimoto T, Renedo M, Wagstaff J et al (1994) Molecular and clinical study of 61 Angelman syndrome patients. Am J Med Genet 52:158–163

Samaco RC, Hogart A, LaSalle JM (2005) Epigenetic overlap in autism-spectrum neurodevelopmental disorders: MECP2 deficiency causes reduced expression of UBE3A and GABRB3. Hum Mol Genet 14:483–492

Shapira NA, Lessig MC, Murphy TK, Driscoll DJ, Goodman WK (2002) Topiramate attenuates self-injurious behaviour in Prader-Willi Syndrome. Int J Neuropsychopharmacol 5:141–145

Sherr EH (2003) The ARX story (epilepsy, mental retardation, autism, and cerebral malformations): one gene leads to many phenotypes. Curr Opin Pediatr 15:567–571

Shields WD (2004) Surgical treatment of refractory epilepsy. Curr Treat Options Neurol 6:349–356

Smalley SL (1998) Autism and tuberous sclerosis. J Autism Dev Disord 28:407–414

Smathers SA, Wilson JG, Nigro MA (2003) Topiramate effectiveness in Prader-Willi syndrome. Pediatr Neurol 28:130–133

Szabat E, Soinila S, Happola O, Linnala A, Virtanen I (1992) A new monoclonal antibody against the GABA-protein conjugate shows immunoreactivity in sensory neurons of the rat. Neuroscience 47:409–420

Tamamaki N, Fujimori KE, Takauji R (1997) Origin and route of tangentially migrating neurons in the developing neocortical intermediate zone. J Neurosci 17:8313–8323

Tataranni PA, Gautier JF, Chen K, Uecker A, Bandy D, Salbe AD, Pratley RE, Lawson M, Reiman EM, Ravussin E (1999) Neuroanatomical correlates of hunger and satiation in humans using positron emission tomography. Proc Natl Acad Sci USA 96:4569–4574

Tecchio F, Benassi F, Zappasodi F, Gialloreti LE, Palermo M, Seri S, Rossini PM (2003) Auditory sensory processing in autism: a magnetoencephalographic study. Biol Psychiatry 54:647–654

Tomchek SD, Dunn W (2007) Sensory processing in children with and without autism: a comparative study using the short sensory profile. Am J Occup Ther 61:190–200

Tommerdahl M, Tannan V, Cascio CJ, Baranek GT, Whitsel BL (2007) Vibrotactile adaptation fails to enhance spatial localization in adults with autism. Brain Res 1154:116–123

Tuchman R, Rapin I (2002) Epilepsy in autism. Lancet Neurol 1:352–358

Valencia I, Legido A, Yelin K, Khurana D, Kothare SV, Katsetos CD (2006) Anomalous inhibitory circuits in cortical tubers of human tuberous sclerosis complex associated with refractory epilepsy: aberrant expression of parvalbumin and calbindin-D28k in dysplastic cortex. J Child Neurol 21:1058–1063

Valente KD, Koiffmann CP, Fridman C, Varella M, Kok F, Andrade JQ, Grossmann RM, Marques-Dias MJ (2006) Epilepsy in patients with angelman syndrome caused by deletion of the chromosome 15q11–13. Arch Neurol 63:122–128

van Slegtenhorst M, de Hoogt R, Hermans C, Nellist M, Janssen B, Verhoef S, Lindhout D, van den Ouweland A, Halley D, Young J, Burley M, Jeremiah S, Woodward K, Nahmias J, Fox M, Ekong R, Osborne J, Wolfe J, Povey S, Snell RG, Cheadle JP, Jones AC, Tachataki M, Ravine D, Sampson JR, Reeve MP, Richardson P, Wilmer F, Munro C, Hawkins TL, Sepp T, Ali JB, Ward S, Green AJ, Yates JR, Kwiatkowska J, Henske EP, Short MP, Haines JH, Jozwiak S, Kwiatkowski DJ (1997) Identification of the tuberous sclerosis gene TSC1 on chromosome 9q34. Science 277:805–808

Verkerk AJ, Pieretti M, Sutcliffe JS, Fu YH, Kuhl DP, Pizzuti A, Reiner O, Richards S, Victoria MF, Zhang FP et al (1991) Identification of a gene (FMR-1) containing a CGG repeat coincident with a breakpoint cluster region exhibiting length variation in fragile X syndrome. Cell 65:905–914

Vincent JB, Horike SI, Choufani S, Paterson AD, Roberts W, Szatmari P, Weksberg R, Fernandez B, Scherer SW (2006) An inversion inv(4)(p12–p15.3) in autistic siblings implicates the 4p GABA receptor gene cluster. J Med Genet 43:429–434

Volk DW, Austin MC, Pierri JN, Sampson AR, Lewis DA (2000) Decreased glutamic acid decarboxylase67 messenger RNA expression in a subset of prefrontal cortical gamma-aminobutyric acid neurons in subjects with schizophrenia. Arch Gen Psychiatry 57:237–245

Volk D, Austin M, Pierri J, Sampson A, Lewis D (2001) GABA transporter-1 mRNA in the prefrontal cortex in schizophrenia: decreased expression in a subset of neurons. Am J Psychiatry 158:256–265

Volk DW, Pierri JN, Fritschy JM, Auh S, Sampson AR, Lewis DA (2002) Reciprocal alterations in pre- and postsynaptic inhibitory markers at chandelier cell inputs to pyramidal neurons in schizophrenia. Cereb Cortex 12:1063–1070

White R, Hua Y, Scheithauer B, Lynch DR, Henske EP, Crino PB (2001) Selective alterations in glutamate and GABA receptor subunit mRNA expression in dysplastic neurons and giant cells of cortical tubers. Ann Neurol 49:67–78

Williams DL, Goldstein G, Minshew NJ (2006) Neuropsychologic functioning in children with autism: further evidence for disordered complex information-processing. Child Neuropsychol 12:279–298

Wong V (2006) Study of the relationship between tuberous sclerosis complex and autistic disorder. J Child Neurol 21:199–204

Woo TU, Whitehead RE, Melchitzky DS, Lewis DA (1998) A subclass of prefrontal gamma-aminobutyric acid axon terminals are selectively altered in schizophrenia. Proc Natl Acad Sci USA 95:5341–5346

Zafeiriou DI, Ververi A, Vargiami E (2007) Childhood autism and associated comorbidities. Brain Dev 29:257–272

Index

A
Adult plasticity and aging, 7–8
Agmon, A., 93, 98
Alger, B.E., 137, 144
Angelman syndrome (AS), 175–176
Aoki, C., 43
2-Arachidonyl glycerol (2-AG), 138
Auditory and visual system
    experience-dependent plasticity, 79
    experience requirement
        high-frequency inhibition (HFI), 79–80
        low-frequency inhibition (LFI), 80–81
        maintenance *vs.* refinement, 85
    FM rate and direction selectivity, 78–79
    inhibitory plasticity
        homeostatic balance, 83–84
        superior colliculus (SC), hamster, 72
        synaptic mechanisms, inhibition strength and timing, 84–85
        vocalization selectivity, 76–77
    retinocollicular convergence modifying effect
        chronic postnatal NMDAR blockade, 75
        NMDA receptor (NMDAR)-activity, 74
        velocity tuning, 74–75
    sideband inhibition and direction selectivity, adults, 78
    surround inhibition
        chronic NMDAR blockade, 76
        receptive field (RF) properties across sensory systems, 81–82
        velocity tuning process, 72–74
Auditory cortex (ACx)
    chronological perspective, 46
    cochlear damage effect, 45–46
    deafness, impact, 54
    EM-immunocytochemistry, 56
    $GABA_A$ receptors, hearing loss effects, 55
    hearing impairments, 45
    inhibitory synaptic plasticity, 53
    thalamic stimulation, 53
Autism spectrum disorder (ASD), 173–174

B
Barrel cortex, postnatal maturation and plasticity
    experience-dependent plasticity
        GABA and GAD role, 106
        metabotropic and ionotropic glutamate receptors, 104–105
        transcriptional factors and maturation, 105–106
    GABAergic circuits
        activity-independent maturation and plasticity, 103–104
        experience-dependent plasticity, 100–103
    interneuron electrical property
        dendritic gap junction (GJ) coupling process, 95
        experience-dependent maturation, inhibitory interneurons, 95–96
        FS and RS-type firing phenotypes, 93–95
        LTS firing phenotypes, 95
        neocortical interneurons, 92–93
    intracortical inhibitory synaptic transmission
        GABA system, 96–97
        late postnatal and experience-dependent maturation, 98–99
        sensory feed-forward inhibition, interneurons, 99
Beaulieu, C., 101
Ben-Ari, Y., 121

Index

Brain-based disorders
   autism spectrum disorder (ASD), 173–174
   epilepsy, 169–172
   schizophrenia, 172

C
Calretinin (CR)
   caudal ganglionic eminence (CGE), 15–16
   lateral ganglionic eminence (LGE), 16
   medial ganglionic eminence (MGE), 14–15
   rostral migratory stream (RMS), 16–17
Carlson, G.C., 150
Castillo, P.E., 151
Caudal ganglionic eminence (CGE), 15–16
Cerebral cortex
   septal region, 17
   autism spectrum disorder (ASD), 173–174
   tuberous sclerosis (TSC), 174
   intellectual disability, 177
Chevaleyre, V., 151
Connors, B.W., 95, 99
Cortical interneurons
   calretinin (CR), 15
   GABAergic interneuronal progenitors, 18
   Nkx2.1, 19
   origin
      caudal ganglionic eminence (CGE), 15–16
      cortex, 17–18
      lateral ganglionic eminence (LGE), 16
      medial ganglionic eminence (MGE), 14–15
      rostral migratory stream (RMS), 16–17
      septal region, 17
      sonic hedgehog (Shh) signaling, 19
Co-transmitters, 5
Critical period
   ocular dominance plasticity, 7
   GAD65 expression, 98, 100
   whisker stimulation, 103

D
Dark rearing
   surround inhibition, 83
   maturation, GABA transmission, 104
Deafferentation
   MNTB neurons, 49–50
   sound-evoked activity, 61
Depolarization-induced suppression of inhibition (DSI), 5–6
   endogenous cannabinoids (eCBs), 139–140
   glutamatergic terminals, 145–146

Diana, M.A., 148
Direction selectivity
   experience requirement, 80–81
   frequency-modulation (FM) rate, 78–79
   sideband inhibition, 78

E
Eagleson, K.L., 167
Edwards, D.A., 149
Elphick, M.R., 139
Endocannabinoids. *See* Endogenous cannabinoids (eCBs)
Endogenous cannabinoids (eCBs)
   and development, 156–157
   history and pharmacology
      AEA and 2-AG, 138
      CB1R, 137–138
      delta-9 tetrahydrocannabinol (THC), 137
      ubiquity, 138
   neurophysiology
      2-AG, 144–145
      assaying methods, 142
      de novo postsynaptic synthesis, 142–143
      depolarization-induced suppression of inhibition (DSI), 139–140
      eCBs and brain development, 146–147
      glutamatergic terminals, 145–146
      GPCR-dependent eCB mobilization, 140–141
      interneurons release eCBs, 147–148
      mobilization, 143
      pre-endocannabinoid, 143
      retrograde signaling, 139
      timing, eCB mobilization, 143–144
   nitric oxide system, synergy, 157–158
   and synaptic plasticity neurophysiology
      DSI and long-term potentiation (LTP), 149–150
      exogenous cannabinoids relationship, 154
      inhibitory long-term depression (iLTD), 150–154
      seizures, 155–156
      spike-timing dependent plasticity (STDP), 154–155
      use-dependent regulation of, 148–149
Epilepsy, 169–172
Ergetova, M., 139
Experience-dependent maturation
   inhibitory interneurons, electrophysiological property, 95–96
   postnatal maturation, 98–99

Index

postsynaptic maturation, 98
presynaptic maturation, 98
Experience-dependent plasticity
    GABA and GAD role, 106
    metabotropic and ionotropic glutamate receptors, 104–105
    transcriptional factors and maturation, 105–106

F
Fragile X, 174–175
Fuchs, J.L., 103
Fuzessery, Z.M., 71

G
GABAergic system
    hippocampal development
        ion transport and E_{GABA} control, 119–120
        tonic actions, 117–118
        trophic actions, 118
    interneuron dysfunction, 170–171
    synapses maturation
        AMPAergic synapses, 33
        correlation-based [Hebbian] rules, 32–33
        early embryos, 35–36
        embryonic limb movements, 29–30
        episode and NKCC1 cotransporter modulation, 31
        function, 27
        $GABA_A$ agonist and voltage-gated channel blockage, 35
        homeostatic plasticity, 33
        inter-episode intervals (IEI), 31–32
        intracellular chloride concentration, 32
        lidocaine-treated embryo, 35
        membrane depolarization, 28
        miniature postsynaptic current (mPSC) amplitude, 33–34
        network-induced depression, 30–31
        neural circuits formation, 33–34
        neuronal activity, 28
        spontaneous network activity (SNA), 28–29
        synaptic strength regulation, 32–33
Ganglionic eminence
    caudal, 15–16
    lateral, 16
    medial, 14–15
Gene regulatory disorders
    Angelman syndrome (AS), 175–176
    intellectual disability (ID), 177
    Prader–Willi syndrome (PWS), 175–176
Giant depolarizing potentials (GDPs)
    acronym, 122
    pacemaker region (CA3), 122–123
    synaptic and cellular mechanisms, generation
        CA3 pyramidal neurons intrinsic bursting, 125–127
        conditional pacemakers, CA3, 127–128
        GABA action developmental shift, 123–125
        glutamatergic transmission, 125
Glorioso, C., 106
Glutamatergic transmission, 125
Gonzalez-Islas, C.E., 27

H
Hammock, E.A.D., 167
Hebb, D.O., 3
Heinbockel, T., 143
Hensch, T., 7
Hibbard, L.S., 103
Hippocampal development
    GABAergic transmission
        ion transport and E_{GABA} control, 118–120
        tonic actions, 117–118
        trophic actions, 118
    giant depolarizing potentials (GDPs)
        acronym, 122
        pacemaker region (CA3), 122–123
        synaptic and cellular mechanisms, generation, 123–128
    ontogeny, 120–122
Homeostatic plasticity, 6–7, 83–84
Hubel, D.H., 71
Hubner, C.A., 119
Huxley, T.H., 115

I
Inferior colliculus (IC)
    chloride (Cl-), 50–52
    GABA, 52
    inhibitory postsynaptic potentials, 50
    synaptic plasticity, inhibitory, 51
Inhibitory gain
    regulatory mechanisms
        IPSC amplitude, 59–60
        kinase-dependent phosphorylation, 59
    auditory processing, 60–62

Inhibitory plasticity
  adult plasticity and aging, 7–8
  co-transmitters, 5
  depolarization-induced suppression
      of inhibition (DSI), 5–6
  homeostatic, 6–7
  ocular dominance, 7
  receptor subunit composition, 5
  receptor trafficking, 6
  spike timing-dependent plasticity
      (STDP), 6
Inhibitory synapse regulation, auditory CNS
    development
  activity-dependent plasticity, 43
  auditory cortex (ACx)
    chronological perspective, 46
    cochlear damage effect, 45–46
    deafness, impact, 54
    EM-immunocytochemistry, 56
    $GABA_A$ receptors, hearing loss
        effects, 55
    hearing impairments, 45
    inhibitory synaptic plasticity, 53
    thalamic stimulation, 53
  cell adhesion molecules, 43
  cellular mechanisms, 59–60
  GABA, hearing loss effect, 57
  heirarchical modification, 58–59
  inferior colliculus (IC)
    chloride (Cl^-), 50–52
    GABA, 52
    inhibitory postsynaptic potentials, 50
    synaptic plasticity, inhibitory, 51
  in vivo manipulations, 61–62
  lateral superior olivary (LSO) neurons
    anti-homeostatic mechanism,
        49–50
    excitatory synapses, 48
    $GABA_B$ receptor, 49
    inhibitory synaptic plasticity, 47
    interaural level differences
        (ILD), 46
    LTD, inhibitory, 48
    medial nucleus of the trapezoid body
        (MNTB), 48
  presbycusis, 61
  spontaneous and sound-evoked activity
    action potential, 44
    neural activity, 44
    retinal activity, 45
  unilateral deafferentation, 61
Intellectual disability (ID), 177
Inter-episode intervals (IEI), 29, 31–32

Issac, J.T., 99
Itami, C., 104

J
Jean-Xavier, C., 124
Jones, E.G., 96

K
Kaeser, P.S., 153
Kaila, K., 115
Kathie, L.E., 167
KCC2. See K^+-Cl^- cotransporter
K^+-Cl^- cotransporter (KCC2), 50,
    119–120
Kim, J., 144
Knott, G.W., 103
Knudsen, E.I., 82
Kotak, V.C., 43
Kreitzer, A.C., 145

L
Land, P.W., 95, 98
Lateral ganglionic eminence (LGE), 16
Lateral superior olivary (LSO) neurons
  anti-homeostatic mechanism, 49–50
  excitatory synapses, 48
  $GABA_B$ receptor, 49
  inhibitory synaptic plasticity, 47
  interaural level differences (ILD), 46
  LTD, inhibitory, 48
  medial nucleus of the trapezoid body
      (MNTB), 48
Levitt, P., 8, 167
Long-term depression (LTD)
  endocannabinoids (ecs), 5–6
  inhibitory plasticity (LTDi)
    DSI-induced LTD, 150–151
    long-lasting and CB1R-dependent
        facilitatory effect, 152
    eCB effects, 153
    cAMP/PKA signaling, 153–154
  inhibitory gain, 59–60
  spike-timing dependent plasticity,
      154–155
Long-term potentiation (LTP)
  DSI, 149–150
  inhibitory plasticity (LTPi), 5–6
  receptor trafficking, 6
  spike-timing dependent plasticity (STDP),
      154–155

Index

Low-threshold spiking (LTS)
  eCB release, 148
  electrical properties, 92–93

M
Marty, A., 148
Masking
  backward and forward, 73–74
  surround inhibition, 75–76
  sideband inhibition, 78
Massengill, J.L., 93
McCasland, J.S., 98, 103
Medial ganglionic eminence (MGE)
  interneuron migration, 21
  Lhx6, 20
  Nkx2.1 and Nkx6.2, 20
  in utero transplantation, 22
Metabotropic glutamate receptors (mGluRs), 105
Micheva, K.D., 101
Minshew, N.J., 173
Mower, G.D., 97

N
Na^+-K^+-Cl^- cotransporter (NKCC1)
  GABAergic transmission, 28
  hippocampal neuron regulation, 119
  inferior colliculus (IC), 50
N-arachidonyl ethanolamine (AEA), 138
Nedivi, E., 8
Nitric oxide system, 157–158
NKCC1. See Na^+-K^+-Cl^- cotransporter
NMDA receptor (NMDARs), 104–105

P
Pallas, S.L., 71
Pathophysiology, interneuron
  brain-based disorders
    autism spectrum disorder (ASD), 173–174
    epilepsy, 169–172
    schizophrenia, 172
    fragile X, 174–175
  gene regulatory disorders
    Angelman syndrome (AS), 175–176
    intellectual disability (ID), 177
    Prader–Willi syndrome (PWS), 175–176
    tuberous sclerosis (TSC), 174
Plasticity. See Developmental plasticity, auditory system

Poo, M.M., 6
Prader–Willi syndrome (PWS), 175–176
Prince, D., 93

R
Razak, K.A., 71
Receptive field properties. See Auditory and visual Systems
Receptor trafficking, 6
Regehr, W.G., 145
Reich, C.G., 150
Response selectivity, inhibitory plasticity, 82–83
Retinocollicular
  inhibitory plasticity, 72
  surround inhibition, 74–75
Retinotectal
  inhibitory plasticity, 72
  N-methyl-D-aspartate receptors (NMDARs), 104–105
Rostral migratory stream (RMS), 16–17

S
Salazar, E., 103
Sanes, D.H., 9, 43
Sarro, E.C., 43
Schizophrenia, 172
Sensory deprivation
  in vitro electrophysiological and neuroanatomical studies, 101–102
  in vivo electrophysiological studies, 100–101
Sideband inhibition, 78
Simons, D.J., 95, 96, 98, 100, 101
Sipilä, S.T., 115
Somatosensory cortex
  rostral migratory stream (RMS), 16–17
  response selectivity development, 82–83
Sonic hedgehog (Shh) signaling, 19
Spike-timing dependent plasticity (STDP)
  endogenous cannabinoids (eCBs), 154–155
  FM direction selectivity, 85
  inhibitory plasticity, 6
Spontaneous network activity (SNA), 27
Suga, N., 78
Sun, Q.-Q., 91, 98
Superior colliculus (SC)
  inhibitory plasticity, 72
  sound-evoked activity, 44–45

Surround inhibition
  chronic NMDAR blockade, 76
  retinocollicular convergence effect, 74–75
  velocity tuning, SC, 72–74
Swadlow, H.A., 99

T
Tuberous sclerosis (TSC), 174
Takesian, A.E.

V
Velocity tuning, SC, 72–74, 76
Vocalization selectivity, auditory system, 76–77

Visual system. *See also* Auditory and visual Systems
  surround inhibition, 72–74
  homeostatic plasticity of inhibition, 83–84

W
Welker, E., 103
Wenner, P., 27
Wiesel, T.N., 71
Woolsey, T.A., 98
Wu, G.K., 84

Z
Zheng, W., 82